UNEARTHING
THE UNDERWORLD

UNEARTHING
THE UNDERWORLD

A Natural History
of Rocks

Ken McNamara

REAKTION BOOKS

Published by Reaktion Books Ltd
Unit 32, Waterside
44–48 Wharf Road
London N1 7UX, UK
www.reaktionbooks.co.uk

First published 2023
Copyright © Ken McNamara 2023

Printed and bound in India by Replika Press Pvt. Ltd

A catalogue record for this book is available from the British Library

ISBN 978 1 78914 718 6

CONTENTS

ONE

EARTH'S DIRTY LITTLE SECRETS

. . . and these old stones, give them a voice
and what a tale they'd tell.

Aeschylus

People have had a long and often very fraught relationship with rocks.
They have used them and abused them for thousands of years: blown
them up, used them to barricade against marauding hordes, built
homes and places of worship out of them. They have even, on a block of
400-million-year-old sandstone, crowned kings and queens. Some rocks,
contentiously, we turn to for heat and light, while others, by the strange
alchemy of mixing two together, have provided us with motor cars and
baked bean cans. Rocks have also long given us the raw material with
which to express our artistic abilities as well as facilitated the rise of many
civilizations. But they are still bloody annoying when they take
up residence in our shoes.

The Colour of Time

Some were the colour of night. Others as white as an egret's wing. Or yellow, like the yolk of the egg that John Michell had eaten for breakfast. Had he stopped his carriage, which rattled through the English countryside, to examine the rocks, Michell might have been struck not only by their colour, but by how different they each felt – smooth and silky, gritty or as crumbly as raw sugar. But Michell was in a hurry. He had no time to stop. He had no time to stare. This was for the simple reason that he was bound for his weekly London meal. Yet Michell did notice the rocks as they rattled past. Indeed, he probably noticed them more than anybody else had before him.

In the 1760s John Michell (1724–1793) was a frequent attendee at weekly meetings of the Royal Society, following which dinner was had either at the Crown and Anchor Tavern in the Strand or at the Cat and Bagpipes, a combined public house and chop-house on the corner of Downing Street. Michell, all agreed, was no Adonis. Clergyman and antiquary William Cole observed that he was a 'little short Man, of a black Complexion, and fat'. Nevertheless 'he was esteemed a very ingenious Man and an excellent Philosopher.'[1] This must rank as one of history's greatest understatements for a man whom few today have heard of, but who was arguably one of the most brilliant and original thinkers of his day. He certainly had an interest in geology, having been the first to suggest that earthquakes travelled in waves. In astronomy this Fellow of Queens' College, Cambridge, was literally centuries ahead of time, arguing for the existence of binary stars and even postulating the existence of black holes. The gravitational pull of some stars was so strong, he argued, that even light would not be able to escape them. For good measure he came up with a way to measure the mass of Earth, and was the first to propose how magnets could be artificially created.[2]

Yet among these mighty thoughts it was his more humble observations on the rocks that he saw on his many journeys between London and Yorkshire, where he was a rector, which reveal some of the first glimmers of a quest to understand the rocks that lay beneath the pastures and fields that clothed the land. The roads on which Michell travelled, whether dry or muddy, would have been either white and chalky, red or grey and muddy, yellow and dusty, depending on the geology of this underworld. Michell realized that the roads upon which his carriage was running intercepted each of these strata in stately succession, the layers of rocks having been gently tilted, one lying on top of the other, like pages in a book:

> This situation of the strata may be not unaptly represented in the following manner. Let a number of leaves of paper, of several different sorts or colours, be pasted upon one another; then bending them up together into a ridge in the middle, conceive them to be reduced again to a level surface, by a plane so passing through them, as to cut off all the part that had been raised; ... this will be a good general representation of most, if not all, large tracts of mountainous countries ... throughout the whole world.[3]

This he wrote in 1759 in the *Philosophical Transactions of the Royal Society*, shortly before becoming Woodwardian Professor of Geology at the University of Cambridge. Moreover, Earth

> (as far as we can judge from the appearances) is not composed of heaps of matter casually thrown together, but of regular and uniform strata. These strata, though they frequently do not exceed a few feet, or perhaps a few inches in thickness, yet often extend in length and breadth for many miles, and this without varying their thickness considerably. The same stratum also preserves a uniform character throughout, though the strata immediately next to each other are often totally different.[4]

Largely forgotten today is a short note written by Michell's friend John Smeaton (1724–1792) in 1788. Not quite on the back of an

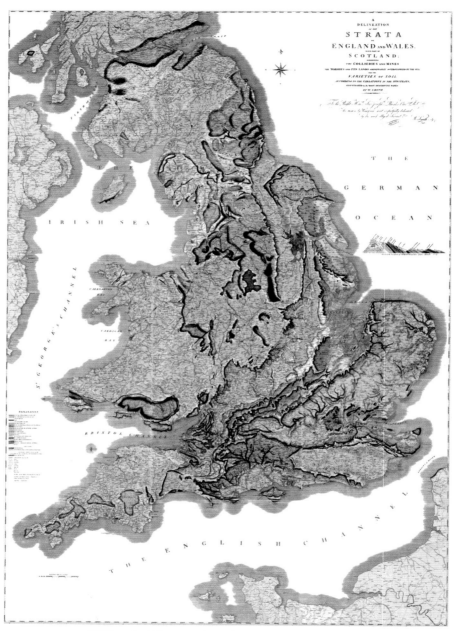

William Smith's 1815 geological map of England and Wales.

envelope, but on the back of a letter postmarked 21 November 1788, Smeaton wrote down Michell's verbal account of the sequence of rocks that lay across the southern and eastern parts of England, in correct stratigraphic succession, and with postulated thicknesses of the various strata. The note was discovered in 1810 among Smeaton's papers, which were then in the possession of Sir Joseph Banks.[5] From the lowest units, the black 'Coal strata of Yorkshire' through the 'Red clay of Tuxford', the blue-grey 'Lyas strata', the yellow 'Northamptonshire lime' to the overlying dark grey 'Golt' (today's Gault Clay) and white 'Chalk', Michell correctly described the essential sequence of late Palaeozoic and Mesozoic strata of much of the Midlands and southeast England. In other words, the rocks that we now know had originally been deposited as sediments between about 300 million and 70 million years ago. How these rocks had formed, though, was a mystery. Realizing that there was some sort of order to their layers was, at the time, quite enough.

Like Michell, William Smith (1769–1839) was also intrigued by the colours of rocks. In the general scheme of our understanding of the modern history of geology (in other words, the history of rocks), much seems to start with William Smith at the beginning of

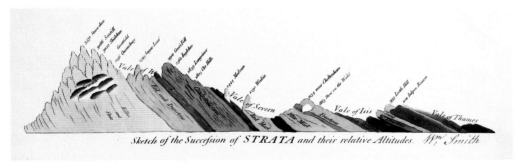

Sketch of the Succession of STRATA and their relative Altitudes. W^m. Smith

William Smith's 1815 *Sketch of the Succession of Strata and Their Relative Altitudes*, being a geological section from Snowdon to the 'Vale of Thames'.

the nineteenth century. A canal surveyor and geological map-maker extraordinaire, Smith is generally acknowledged as the man who brought rocks into the daylight and began the quest for a rational understanding of how they were made and what stories they had to tell. He also realized that it was possible to depict these rocks from the fields and hills on sheets of paper, producing in 1815 the first comprehensive geological map of England and Wales.[6] Smith's masterstroke was to colour in the different rock types on his map in much the same hue as they existed in nature: yellow for the yellow Jurassic limestones; red for the red marls and sandstones of the Triassic; blue for the early Jurassic 'Blue Lias'. Chalk, though, was green, not white, maybe because chalk downland is often covered by green grassy swards. Differences in colour were the clearest way of differentiating between different rock types. However, what caused these differences in colour, and what they meant in terms of the formation of the rock, were questions that, it seems, hardly impinged upon the consciousness of these early geological pioneers.

Smeaton's doodling on the back of a letter shows that Smith did not appear from a dark geological void. There had been others, like Michell, now mostly forgotten characters in the history of science, who even as early as the late seventeenth century were both intrigued and perplexed by the common or garden rocks that lay beneath their feet. One man, though, stands out as having had an almost pathological fixation with rocks, long before anyone else took any serious notice of them. His name was John Woodward.

Reading the Rocks

It is 1688 and John Woodward (1667–1728), a 21-year-old fledgling physician, is rummaging in a gravel pit among some new buildings under construction in Dover Street, St James's, in London. A small pebble, one among thousands, attracts his attention, so he picks it up. As he was to write forty years later, it was 'the first stone I ever took notice of, or gather'd'. It turned out to be the first of many thousands of specimens that he collected, transforming Woodward into one of the leading geologists of his time (though the word 'geologist' had not then seen the light of day). Today this little pebble sits in the University of Cambridge's Sedgwick Museum with nearly 10,000 other rocks, minerals and fossils that Woodward subsequently collected. There is nothing special about this first one. Rather, as Woodward noted, it's just a rather dull *gritty Peble*. But when he wrote up the catalogue of his collection many years later, he did what nobody had ever bothered to do before. He proceeded to describe his 'Peble' in inordinate detail.

His 'Peble' he also calls a 'stone', a term that he sometimes uses interchangeably with 'rock'. Yet what, if anything, is the difference between the two? Searching for simple, straightforward definitions of these two terms is somewhat akin to holding a handful of fine sand. One second you think you have a firm grip on it; the next it drifts through your fingers and is lost. John Woodward, often perhaps subconsciously, sometimes uses the term 'rock' when describing a large mass that forms a cliff or mountain. As the *Oxford English Dictionary* would have it, a rock is 'a large, ragged mass of hard mineral material or stone forming a cliff, crag or other natural feature on land or in the sea'. When talking about a small object that he is holding in his hand and is part of his collection, Woodward sometimes still calls it a rock; other times it becomes a 'stone'. To the OED a stone is 'a piece of rock . . . of a small or moderate size'. If smooth, having been abraded by water, ice or wind, it becomes a pebble, such as he collected in Dover Street. This term encapsulates within it more of the history of the object. It has been eroded, rolled around, worn, transported and smoothed by abrasion with its fellow travellers. These days every student of geology has it drummed into them that rocks are rocks and stones are rocks. And so it will be. It is hard to shake off old habits.

Woodward realized that *every* rock (be it a pebble, stone or rock) is important and has its own tale to tell, however nondescript it may appear. He dips his pen in ink and begins to write:

> A gritty *Peble* of a very light brown Colour, an oblong oval shape, an Inch and 3/4 in length, and one Inch in breadth, flattish, and having the two Ends somewhat pointed. There's a narrow ridge, of the same breadth in all parts, running directly long-ways of the Stone, and quite encompassing it. This Ridge consists of a closer and harder sort of Matter than the rest of the Stone.[7]

While a picture might paint a thousand words, here Woodward is doing something no other writer had ever done before – trying to present a picture of the stone to the reader in, usually, less than a thousand words. His catalogue, like his book on the Earth's history, carries no illustrations. So Woodward asks the reader to exercise their imagination and create a picture of the 'Peble' in their mind. But he doesn't stop there. He wants to explain to the reader why it has the shape that it has. How can he do this? Were not all rocks created by God? Who is he, a mere mortal, to think that he can question God's intentions? This is encroaching on the heretical. He has no prior knowledge of this sort of rock, nor of rock-forming processes. At this time, late in the seventeenth century, nobody has. But Woodward is made of stern stuff. Some of his contemporaries might say, overtly arrogant stuff. What he proceeds to do is to apply simple, analytical, empirical logic to try to understand how the rock formed. He wants to read the rock. He continues writing:

> In the middle on one side, the Stone sinks in, and rises out on the opposite, as if it had been soft and press'd in that Part. Indeed it appears, upon the whole, as if it had been flat the quite contrary way of what it is at present . . . till some exterior Force compressing the two opposite Edges, brought it to the Form it now contains, which indeed is very odd and extraordinary . . . 'Tis not likely that the Ridge was extant at the first Formation of the Stone but it appears very naturally as if it had been soft, and then compress'd, and wrought into the Shape it now bears.[8]

Time is at work. To Woodward's mind, some unknown force is moulding and squeezing the rock into its final shape. It had been soft but now it is hard. Woodward gives his rock its own number in

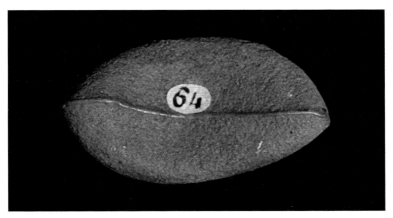

A worn sandstone pebble found by John Woodward in Dover Street, St James's, London, in 1688. As Woodward wrote forty years later, it is 'the first stone I ever took notice of, or gather'd'.

his collection – c.226. But it is with the description of the next stone in his catalogue, number c.227, that Woodward really goes to town: two pages of detailed description and analysis of a pebble in which he travels back into an unknown past (to him it was during the biblical Flood), invoking a range of geological, erosive processes that could have worked on this battered piece of flint, and then coming to his conclusion, which forms the basis for his much-derided history of Earth, published in 1695 – that the present form of such rocks was due to having been caught up in the 'Waters of the Deluge'.

Stones that are tumbled about upon the shores of the Sea, agitated by Tides and Storms . . . these bodies have had their surfaces ground, and worn . . . the Force of Water . . . was

One of John Woodward's walnut cabinets for his rock collection, now housed in the Sedgwick Museum, Cambridge.

so great, as to tear up some of even the most solid Strata, tumbling them along, rounding and smoothing them.[9]

Nobody, *nobody*, had ever peered this deeply, and with such imagination, into a single rock.

It doesn't matter how attractive, weird or 'curious' the rock is. Every stone has something to say about how Earth came into being. And by the simple act of wondering how this rock had formed, Woodward transforms the collecting of rocks, fossils and other subterranean lithological fancies from a pastime for idle gentlemen into a science. In one respect, Woodward gets it right. He realizes that to understand Earth's formation, it is necessary to carry out careful observations and documentation of as many different types of 'fossil' as possible, from all over the world. To him, and his contemporaries, all rocks, minerals and what we now recognize as fossils fell under the umbrella term of 'fossil', derived from the Latin word *fossilis*, meaning 'dug up'. In his pebbles he sees the effect of erosion: the pummelling of strata by the sea, the transport of the pebbles and grains of sand, is a process that takes place today but which also occurred in just the same manner in the distant past. He realizes that these rocks have their own history, which can only be discovered by learning the rocks' language. Many of the observations and interpretations that he made in his two catalogues, published after his death in 1728 and 1729, are hundreds of years ahead of their time. Yet nobody bothers with them. They are small fry, lurking deep in the obscurity of his painstakingly constructed records.

While Woodward attempts to interpret his rocks in this novel, scientific way, his deductions, when viewed through a three-hundred-year-old lens, are often, with the benefit of hindsight, spectacularly wrong. Like all of his contemporaries, he had no concept of the depth of time. If you had told him that Earth had formed 4.6 billion years ago, he would have laughed in your face, or gone out of his way to have you committed to the newly constructed Bethlem Hospital mental asylum. Every one of his precious stones is, to Woodward, proof positive of the hand of God as told in the biblical account of the Flood. He shoehorns all his careful, scientific observations into what almost all thinkers of the time believe: Earth had been devoured in an all-consuming cataclysmic Flood. What he sees in his rocks are the results of the accumulation of the earthly debris following the global carnage. It all happened, he believed, in just a few days.

Yet to Woodward goes the award of being the first man of science to realize that rocks had their own personal stories. They could tell you their life histories if you listen closely enough. This is what he set out to do in the catalogue to his great collection. According to Woodward,

> I should have been glad I could, in each, have also set forth, what I have of several, at what Depth, and in what Manner it lay: among what Matter it was reposited,: [*sic*] as also in what Plenty; along with all the other considerable Circumstances of it. This is follow'd, as far as my other Affairs would permit, by particular Observations upon the body itself: upon the Colour, the Bulk, the Form, the Texture, the Constitution, the Purity of Mixtures and Remark.[10]

'In fine', he continued, 'upon these Histories, Accounts, and Observations, are founded several Reflections and Inferences relating to the Origin and Formation of the Body.' There was no point, in Woodward's view, in just making observations of geological specimens. It was imperative to use them as launching points for a greater

One of the trays from John Woodward's cabinets containing colourful 'Earths and Earthy Substances'.

understanding of how rocks formed and what they can reveal of the ancient worlds from which they came.

> I know well there are Those who would have the Study of Nature restrained wholly to Observations; without ever proceeding further. But due Consideration, and a deeper Insight into Things, would soon have undeceived and made them sensible of their Error. Assuredly, that Man who should spend his whole Life in amassing together Stone, Timber, and other Materials for Building, without ever aiming at the making an Use, or raising any Fabrick out of them, might well be reputed very fantastic and extravagant. And a like Censure would be his Due, who should be perpetually heaping up of Natural Collections, without Design of building a Structure of Philosophy out of them, or advancing some Propositions that might turn to the Benefit and Advantage of the World. This is in reality the true and only proper End of Collections, of Observations, and Natural History: and they are of no manner of Use or Value without it.[11]

The colours of rocks, their textures, their compositions: these are the underlying grammar of the rocks and their natural histories. Woodward recognized this more than three hundred years ago, and the same still holds true today.

* * *

TO BUILD HIS COLLECTION, John Woodward seems to have travelled widely to many parts of England. One such trip was to the Isle of Wight. Today Alum Bay is famous for its multicoloured sands. It was a well-known site in the late seventeenth and early eighteenth centuries, particularly for white sand, used in making crystal glass, as well as clay for making tobacco clay pipes. Woodward collected samples of both. The pipe clay, he thought, 'is very fine. 'Tis indeed the finest in England.'[12] From Alum Bay he also collected 'Sand, of a pale red colour, approaching a Pink', as well as a 'reddish brown Sand'. Not all of the clays he collected were grey pipe clay. Some were 'Yellow ochreous Earth'; others 'Ochreous Earth, of a red Colour, somewhat approaching a Pink'.[13] Exactly where on the island he dug them out he didn't record. However, there is a place nearby on the island where similarly kaleidoscopic clays hold the clue to the environment on that part of the island some 130 million years ago. And it all comes down to colour.

Rocks do not have to be as hard as, well, stone. From a geological perspective, clay that you can knead with your hands still counts as a rock. And on the southern coast of the Isle of Wight there is a lot of it, forever battered by winds and storms that come scudding up the English Channel. Winter, unsurprisingly, is the worst. Where the coastline is made up of rocks of stern constitution, such as well-cemented chalk, then they can, up to a point, withstand the endless battering. But those parts of the coast made merely of clay ceaselessly wash into the sea. Fortunately, this exposes fresh outcrops of clay of many colours – green, grey, red, yellow and purple, each reflecting a subtle variation in environmental conditions at the time of deposition.

When you stand with your back to the freezing, howling winds that scarify the cliffs, it is almost inconceivable to imagine that 100 million years earlier during the early Cretaceous period, this place was a hot, and at times humid, landscape, not unlike the present-day Okavango Delta region in Botswana, the Everglades in Florida or Kakadu in northern Australia. While pods of hippos may not have

Multicoloured clays formed in Cretaceous wetlands on the
Isle of Wight 130 million years ago. This sand layer was punctured
by a dinosaur's foot.

carved channels through these Cretaceous deltaic sediments as they
do in the Okavango Delta, there were crocodiles lurking in the shallow billabongs, much like in Kakadu today. As the sky blackened
with passing flocks of pterosaurs, the ephemeral ponds that formed
on these seasonal tropical wetland floodplains were also visited by
pods of dinosaurs, many leaving their footprints, and sometimes
their bones, in the clay.

The variety of colours in the clays are produced by different
compounds of iron. These differences reflect formation in subtly
changing environments in the shallow pools that dotted an extensive

river floodplain – topographic variations of the floodplains and seasonally variable periods of flooding.[14] Three types of mudstone can
be seen in these crumbling cliffs, and each grades imperceptibly into
the other. Green-grey clays reflect ponds in which water persisted
for long periods and often became stagnant due to very low oxygen
levels. Plant remains often occur in them in the form of black, shiny
coal. They also contain abundant amounts of fossil charcoal (fusain).
This likely indicates billabongs trapping local flood waters that had
transported sediment following wildfires. The iron minerals siderite (iron carbonate) and pyrite (iron sulphide) cause the green-grey
colour. The second type of clays – those mottled red, purple, yellow –
developed in slightly higher areas, in more oxygenated waters where
there was some free drainage. The different colours are the product
of hydrated iron oxides, in particular the mineral goethite, which
is yellow. Last are the pure red mudstones, which formed on even
higher areas that were only occasionally flooded. Consequently, any
iron that was present occurs in the non-hydrated form of iron oxide:
haematite. Even apparently boring piles of clay, it would seem, can
have an interesting and, dare I say, colourful history.

Recycled Rocks

Like so many other things in nature, rocks are traditionally subdi-
vided into three fundamental types. There are those crystalline rocks
that are predominantly the frozen remains of the molten magma
that lies kilometres beneath our feet on this very active planet – the
so-called igneous rocks. Most, like granite, form deep in the crust and
ultimately find their way to the surface from the erosion of the rocks
above them over millions of years. Others, such as basalt and rhyolite,
are rocks instantly frozen from the spume of volcanoes after shafts

of magma have managed to navigate their way to Earth's surface. Basalt forms from low-viscosity, silica-poor magmas that ooze like languid superheated porridge from rents in Earth's crust; rhyolite from explosive events usually associated with more viscous, highly silica-rich magmas. Volcanic bombs from such explosions have been known to land in the most unusual of places. In Woodward's collection there is a small piece of green solidified magma. He described it as, 'A vitrify'd Substance, variegated with dusky and green, flung forth of Vesuvius. Mr. Bembde was present when it was flung forth. It fell upon his Hat.'[15] Mr Bembde, we presume, survived this aerial bombardment.[16] This book will not dwell for long on igneous rocks.

The second fundamental type of rock also forms deep in Earth's crust: metamorphic rocks. These are rocks that have pre-existed in another guise, maybe as a sandstone, a limestone or a granite, and then have been caught up in the slow crawl of the tectonic plates that encrust Earth like a fragmented turtle carapace. Subducted deep into the crust, to be heated and subjected to unimaginable pressures, sandstone transforms into quartzite, limestone into marble, mudstone into anything from slate to schist and gneiss, ultimately in many cases to be completely melted to be reborn as igneous rocks. But this book has little more to say about metamorphic rocks.

What remains is the subject of this book – a host of familiar, and you might think rather dull, rocks that formed on the surface of this restless planet, whether under the sea or on the surface of dry land, but which have, for the most part, been buried to form a significant part of the underworld: sandstone, mudstone, limestone, coal, flint, but also a bewildering array of less familiar, but no less intriguing, rocks, including silcrete, calcrete, microbialite, sarsen stones, pisolite, wackestone and the delightfully named puddingstone. Such rocks are generally called sedimentary rocks. Yet this is a particularly inadequate term for the gallimaufry of fascinating rocks that litter about three-quarters of Earth's outer skin – rocks whose formation is very much influenced by whether it took place surrounded by water (under the sea, in lakes or in rivers) or the atmosphere (on land). There is no denying that there are plenty of rocks that are consolidated sediments in the sense of sand grains cemented together to become sandstone (similarly, as on the Isle of Wight, mud compressed into mudstone). Such rocks are formed directly from the physical and chemical abrasion of pre-existing rocks. These recycled rocks are only some of the types that fall under this vague banner of 'sedimentary rocks'. There are many more. What they all have in common is that their component parts were ultimately derived from older pre-existing material, and their development has been determined by the environmental conditions existing either at the interface of Earth's outer surface and the oceans that cover two-thirds of it, or where it is enveloped by the atmosphere. Understanding parameters like temperature, rainfall, ocean or atmospheric chemistry, particularly oxygen and carbon dioxide levels, and how they influence the creation of so many rock types allows the environmental conditions on Earth, both physical and biological, to be established at the time of their formation.

Sediments, the *Shorter Oxford Dictionary* tells us, consist of 'particulate matter that is carried by water or wind and deposited on the surface of the land or the bottom of a body of water and may in time become consolidated into rock' – sedimentary rock. But is that all they are? While some 'sedimentary' rocks fall under the umbrella of being made of 'particulate matter', many do not. Those that do fall into two groups. One is called 'clastic' rocks (from the Greek *klastos*, meaning 'broken into pieces'). Clastic rocks are mainly silica-based, broken-up fragments of quartz, derived generally from silica-rich igneous and metamorphic rocks, like granite and gneiss, respectively.

However, there is another altogether different sort that, while similarly clastic, has derived its 'particulate matter' not from pre-existing rocks but from the remains of once-living organisms: 'bioclastic' rocks. For the living organisms that inhabit this planet – animals and plants, bacteria and fungi, even viruses – all quietly make an astounding variety of rocks. In many cases the remains of organisms can be the very essence of rocks. Or their skeletal parts, after death, may dissolve and be reborn in another lithological (rock-making) form altogether, often aided by a range of microbes. Material processed by animals in their guts and voided as faeces makes up an astonishing amount of rocks. Accumulations of plant material in the right climatic conditions over millions of years can produce enormous thicknesses of rocks – rocks in which sunshine stored as chemical energy through photosynthesis has been trapped for hundreds of millions of years. Fire can also contribute to rock formation, the burnt remains of forests often appearing in the rock record as fossilized charcoal.

So, while many rocks have been formed by the power of Earth's atmosphere, with water and atmospheric gases eroding older rocks, many, somewhat surprisingly, are the remains of living organisms. Immediately such rocks provide insights into a host of past environmental conditions – climates, rainfall regimes, variations of particular gases, such as oxygen or carbon dioxide, in the atmosphere. They also illuminate the nature of past life: the composition and textures of rocks formed from the broken remains of once-living organisms will vary as their evolutionary histories unfold. Some rocks even hold the secrets of the fundamental processes of evolution.

A 280-million-year-old Permian silty seabed with a preserved echinoderm meadow of starfish and crinoids.

Yet there are some rocks formed at the Earth's surface that did not come from particles of pre-existing rocks or fragments of animals or plants at all. Rather, they have come from the recycling of their dissolved remains that have found their way into the groundwater or oceans and were then re-precipitated as pristine recycled rocks. How this occurs depends on many factors, but various microorganisms play a crucial but unheralded role in such rock formation. Some rocks are so fussy that they will only form in the presence of certain types of fungi. Perhaps even more strange is that a suite of bacteria, working in concert, have the ability to cause the precipitation of minerals and create what are essentially living rocks, forming reefs many kilometres long. But possibly the most surprising of all is the recent discovery that viruses are crucial to this mineralization process. Essentially, every conceivable type of organism has played a role in producing many of the rocks that we see all around us: in the walls of our houses and places of worship; in great European gothic cathedrals and the ancient pyramids of Egypt; on the walls we sit on to eat our lunch; and under our feet, in the flinty gravel of our footpaths. All are made from rocks that we either steadfastly ignore or abuse on a daily basis, which exist courtesy of the living organisms with which we share our planet. Whether derived from the pulverized particles of former rocks or the crushed-up remains of the bodies of animals, plants or other organisms, or the precipitation of minerals whose component chemicals were acquired from the dissolution of once-existing rocks or organisms, all are, in one way or another, recycled rocks. The planet wastes nothing.

Wheel within Wheels

Sandstone, limestone, chalk, ironstone, mudstone, chert. Each and every rock has something to contribute to unearthing the history of the underworld and this planet that we share. Rocks can be composed of everything from oysters to great forests, from bacteria to the remains of mountains, and tell the never-ending story of how environments and ecosystems on this dynamic planet have evolved for billions of years.

Many rocks are climatically sensitive and can be used as signposts to ancient environmental conditions. So, for instance, evaporites, like rock salt and gypsum, and a type of limestone called 'calcrete', are only known to form in arid and semi-arid conditions. Glacially derived rocks form in freezing conditions. Iron- and aluminium-oxide-rich laterites are produced on land by continental weathering under seasonally hot and humid conditions. Another type of limestone, known as oolitic limestone, forms in warm, shallow seas, but where the climate on land is hot and dry. The presence of particular rock types at certain times in geological history provides an insight into Earth's climatic history and its geography. One study applied this model to estimate past ocean temperatures.[17] On the basis of the global distribution of laterites, evaporites, oolites, calcretes and glacial deposits formed during the Cambrian Period, 540–490 million years ago, the study showed that polar ocean temperatures lay in the range of 9°C to 19°C (48°F–66°F), while seas in tropical latitudes were between 30°C and 38°C (86°F–100°F). Such is the utility of rocks to unlock past climatic conditions.

The formation of many of these rocks is driven by forces acting both internally and externally on our planet. Internally, the production of many of the raw materials of recycled rocks is controlled by plate tectonics. Continents have danced elegantly around the planet for billions of years, throwing up mountains and changing the shapes of oceans. Mountains have chemically and physically eroded and the igneous, metamorphic and sedimentary rocks of which they are made have broken down, in part, to form sediment that is transported, then accumulated. Chemically dissolved elements of the rock are repurposed in recrystallized form, for instance as chert or some limestones. Ultimately the sediments will be subducted, metamorphosed and melted, before returning to the surface through volcanic activity. The rock cycle is complete.

As oceans wax and wane in size with the moving continents, the courses of currents change, modifying ocean temperatures, influencing the climate and affecting the production of carbonate and siliceous sediments. Volcanoes extrude gases, especially CO_2, into the atmosphere, further impacting global temperatures and consequently the rate of weathering of rocks on land. More CO_2 is produced during periods of enhanced plate tectonic activity, causing global temperatures to rise. Drawdown of CO_2 out of the atmosphere by photosynthesizing organisms tempers these increases, producing carbonate sediments, and so directly influencing the types and quantity of the rocks that form. The increase in these sediments during periods of high atmospheric CO_2 results in the formation of more limestones. Many will ultimately be thrust into the atmosphere to be eroded, recycling the trapped CO_2 as bicarbonate ions and further contributing to the generation of carbonate sediments. With their subduction, CO_2 is returned to the mantle. Volcanic activity will return it to the atmosphere, completing the carbon cycle.[18] The role of rocks in this is critical.

Cyclical variability in climate – in temperature, light intensity, rainfall, humidity – played out over hundreds of thousands of years, or tens of thousands or even just annual seasonality, can trigger the production of rhythmic cycles of different types of rock. Each has its own story to tell of the environmental conditions under which it formed. Details of the pace and frequency of climate change are encapsulated within many rocks, recording how carbon dioxide levels, rainfall, temperature and extremes of seasonality have varied over millions of years.

Many of these cycles are driven by astronomical forcing. The most obvious of these are day and night. Then there are the annual turns of the seasons – intense in high latitudes, weak equatorially. Less obvious, but by no means less significant, are the roughly decadal sunspot cycles. These cycles can all have an influence on the formation of chemical sediments, even, surprisingly, the diurnal cycles. Perhaps the most significant cycles driving alternations in sediment deposition are the larger scale so-called Milanković cycles. The activity of microbial communities, and the type of weathering and erosion controlling the geochemistry of the oceans in Earth's early history, were heavily influenced by heat and light: the energy output of the Sun. While its energy output has been slowly increasing since very early in Earth's history, this has occurred in an episodic manner, with the amount of energy from the Sun that hits our planet varying over time with cyclical rhythmicity. It is affected by the distance of the Sun from Earth and irregularities in the rotation of Earth on its axis. These are the great astronomical cycles that have had a major impact on the formation of sedimentary rocks through Earth history – the astronomical rock clocks.

More than a hundred years ago the Serbian scientist Milutin Milanković (1879–1958) proposed that Earth's long-term climate was primarily influenced by changes in its position relative to the Sun, resulting in episodic changes in solar-energy input: heat and light, each controlled by energy output from the Sun. The amount of energy hitting Earth is subject to the vagaries of its orbit around the Sun and the irregularities of its rotation on its axis.

The orbit of the Earth–Moon system around the Sun is not exactly circular. Gravitational pull from Saturn and Jupiter in particular causes Earth's orbit around the Sun to vary from near-circular to elliptical. This Milanković cycle is known as 'eccentricity', and it is the orbital variation in distance between Earth and the Sun. At the present time, one of the results of this eccentricity is that summer in the Northern Hemisphere is four and a half days longer than winter. The difference in the amount of energy in the form of heat and light reaching Earth in this eccentricity cycle is appreciable. At its most elliptic there is about 23 per cent more solar radiation reaching Earth annually than when it is furthest away. Eccentricity cycles have two main periodicities: one has an average of about 100,000 years, the other 400,000 years. There are also other eccentricity patterns over million-year-plus timescales. The 100,000-year eccentricity cycle may have been one factor influencing the cycles of glacial and interglacial periods over the last half a million years.

Another Milanković cycle is the angle that Earth's axis of rotation is tilted as we orbit around the Sun. This is known as 'obliquity'. During the last million years the angle has varied between 22.1 and 24.5 degrees. Currently it is 23.4 degrees, and decreasing in a cycle that takes just over 40,000 years. The greater the tilt of the axis, the more extreme the seasons. Larger angles favour melting of ice sheets, as each hemisphere experiences more solar radiation during the summer months. As the tilt is decreasing, the resultant cooler summers allow the build-up of ice sheets at high latitudes.

The third Milanković cycle is 'precession'. This describes Earth's wobbling axis. Think of a slowly rotating, wobbling spinning top:

that's Earth. The wobble arises from tidal influences of the Sun and Moon causing a bulge in Earth's equator, which has a slight impact on the planet's rotation. The direction of the wobble relative to the fixed position of stars is called 'axial precession' and has a cycle of about 25,000 years. It also has the effect of enhancing seasonal contrasts, increasing in one hemisphere while decreasing in the other. Currently

Cyclical layers of limestone and shale in cliffs at Lyme Regis, Dorset, formed, it is said, by the influence of Milanković cycles in early Jurassic times, nearly 200 million years ago.

it is making Southern Hemisphere summers relatively warmer than those in the Northern Hemisphere.[19]

These astronomically driven cycles have indirectly influenced the formation of some sequences of rocks by producing cycles of climate and sea-level changes. This has led to rhythmic variations in sediment production and deposition over vast tracts of time which can even be detected in rocks billions of years old.[20] Exposures of these banded rocks of the underworld, in sea cliffs, the flanks of ravines and gorges, towering mountains and road and rail cuttings, are often hard to miss. They may even, on occasions, impinge slightly on our consciousness – soft, black shale alternating with hard, grey limestone, cycling over and over again, as in the cliffs at Lyme Regis; the searing white cliffs of the Seven Sisters on the Sussex coast, striped every metre or so by thin, black layers of hard flint; and vast mountain ranges of banded iron formation in the Pilbara region of Western Australia, cycles within cycles of bands of ironstone and chert, ripe for extraction and transformation into metallic iron, all so very hard to miss. Just where would we be without rocks?

So, What Have Rocks Ever Done for Us?

It is not unreasonable, I think, to ask this question. And, equally, it would not be unreasonable to answer that rocks have played a pivotal role in the evolution of human societies. It all probably started about 3.3 million years ago, when our early australopithecine ancestors began hitting things with lumps of rock. A million years later, early species of *Homo* had learnt to craft hard pieces of chert and basalt into practical tools. By comparison with the first efforts, this stone technology was a tremendous leap in cognitive development, for they had mastered the art of manufacturing a symmetrical, multipurpose tool. Many were, indeed, works of art. Fossils found half a million years ago in some of the rocks being worked were pleasing to the eye, so the stone knapper crafted the tool around them.[21] In fossils and rocks, humans had found their aesthetic sense. People then discovered that they looked good if they wore a few suitable pieces of attractive rock. Bling had arrived. But rocks were not always used for positive purposes. They also have a darker past, in their transformation into weapons of conflict.

On the positive side, rocks became sites of shelter, and eventually provided the building blocks for permanent habitation. They began to yield useful material: ochre for painting, both of peoples' bodies and of the rock walls that became canvases for their art. As human societies developed in complexity, people came to realize that mixing certain rocks together and heating them could provide a most useful new material: metal. It was also discovered that other rocks, once great forests, when burnt provided warmth and eventually other forms of energy.

Yet despite the intimate part they have played in our lives, most people know precious little about where rocks have come from or how they formed. We have built our lives out of rocks that so intrigued those such as John Woodward, John Michell and William Smith, yet we have failed to listen to what they can tell us about ancient worlds, now long gone. Perhaps now is a good time to hear their stories.

TWO

INTO THE LIGHT

Within every lump of rocks lies the memory of its journey through time
– its own secret history. Most rocks are so mind-bogglingly old that they
contain within them the stories of past worlds – past life, climates and envi-
ronments. To unlock these memories, all that is needed is an understanding
of the language of the rocks: what their colour means, or their texture
and chemistry. Once understood, vistas of an ancient planet are opened
up – worlds of ice, raging floods, strange unbreathable atmospheres or seas
teeming with life. For rocks are the ultimate tool for time-travelling back to
when life first began on Earth, more than 3.5 billion years ago.

Living Rocks

A small lake near the town of Cervantes in Western Australia is
circled by a narrow band of rock. It is so narrow, in fact, that
you can, if you are feeling particularly revolting, spit from one side
to the other without too much effort. Should you be feeling a little
more genteel, then it is almost narrow enough to be rather energetically jumped across. The only downside to this is that you would
end up in the lake. Less energetically, it takes just a few strides to pass
over these crumbly rocks, from low vegetation to lake. A few white,
crusty domes of rock, breakaways from the narrow band, breach the
surface of its water – convenient perching points for myriad seabirds.
Seemingly inconsequential in such a beautiful landscape, this rock's
relationship to the microscopic life that teems unseen in the lake is
one of profound significance, for it has a history that stretches back
to the first struggles of life on Earth, billions of years before now.

This little body of water is called Lake Thetis. It was named after a schooner that belonged to one Joshua William Gregory. With his brother, he undertook the first serious geological explorations of the fledgling colony of Western Australia, and in 1847 he had ventured to this area 250 kilometres (155 mi.) north of the capital, Perth. What better name to bestow on this little lake, thought Gregory, than the name of your boat? Shaped like a 500-metre-long (545 yd) teardrop, the lake seems bottomless; walk into it, and the water barely covers your head. It tastes like the nearby sea. What makes this lake and its embracing rocks so special is that this is the closest we can ever get to experiencing some of the world's first ecosystems. And what makes the microscopic life in this lake so remarkable is that it produces what are, essentially, living rocks. Now that, I agree, is a bit of a scary concept. Certainly, your chances of holding a reasonably civilized conversation with a lump of granite are remote. But many other types of rocks wouldn't exist without the industrious activity of organisms of all types, in this case a select suite of microorganisms that have the useful facility of being able to produce rocks.

The ability of an organism to grow minerals might, on the surface, seem to be a trifle unusual. But it's not. You did it yourself when you were young and now call the resulting mineral 'bones'. The compound that you grew is calcium phosphate. What many of the microorganisms that call Lake Thetis home can do is, rather impressively, induce the precipitation of calcium carbonate in the form of the mineral aragonite. They source calcium bicarbonate dissolved in the lake water, which itself is derived from the sand dunes that surround the lake. The dune sand grains are mainly fragments of broken shells made of calcium carbonate. As acid rain percolates through the dune sands, the shell fragments slowly dissolve and the calcium carbonate is recycled into the groundwater, feeding the lake with the very stuff needed by the microorganisms to make rocks – in this case, limestone.

There are many different types of limestone, often with their own quite distinctive, and delightful, names: chalk, oolite, travertine, calcrete, tufa, coquina, wackestone, packstone, grainstone and many more. Those that rim Lake Thetis, constructed by the activity of microbial communities, are usually called stromatolites. I say 'usually' because in recent times some scientists prefer to call them 'microbialites'. Then there are those who call similar rocks 'thrombolites'. Internally they look like a heavily calcified brain, but they form in much the same way as stromatolites. I will follow the most recent major publication dealing with microbially generated limestones and call them 'microbialites'.[1] The two major types are stromatolites and thrombolites, the former layered, the latter not.

Microbialites growing in saline water, Lake Thetis, Western Australia.

In Lake Thetis the lumps of limestone, so cunningly crafted by microbes, come in many guises. Hence the name of the lake in which they are 'growing' unwittingly seems very appropriate. Thetis was a goddess of the sea, and, in addition to possessing the gift of prophecy, she was able to change her shape at will. This is exactly what these microbial communities have been doing in this lake for the last 4,000 years: constructing at a glacial pace an encircling reef of limestone composed of a visual cacophony of shapes. In some places low, layered domes the size of a small car breach the lake's surface; elsewhere, lumps like white car tyres with a hideous skin disease, but with a solid core of calcium carbonate, morph into branching fingers that point to the sky, seemingly floating on the lake. The solid core is thrombolitic microbialite, the outer, layered part stromatolitic fingers. In many places along the lake edge, microbial communities build, somewhat prosaically, flat platforms of limestone.[2] Scientists are only just coming to grips with the factors that determine these various shapes. The hope is that by understanding what controls the formation of microbialites in modern ecosystems, the secret histories of similar rocks millions of years old can be revealed.

The remarkable ecosystem of microbes that constructs these living rocks is mainly a pot-pourri of bacteria. These are not the sneezing, sore-throat-creating, let's-give-this-human-a-runny-nose-for-two-weeks kinds of bacteria, but ones that possess a miraculous power. It is a power that they evolved billions of years ago, when our Sun was a pale imitation of its present self. It is a power that determines life on Earth, the miraculous ability to turn the Sun's energy into life: photosynthesis. We tend to think of this amazing facility, which we take so much for granted, as being restricted to plants. Not so. Long before plants evolved, a group of bacteria called cyanobacteria evolved the ability to photosynthesize. They used to be called 'blue-green algae' but aren't any more, for the very good reason that they are not algae, but bacteria. Taking carbon dioxide from the atmosphere, water from the oceans and light from the Sun, they synthesized organic compounds, producing oxygen as a by-product. Although cyanobacteria might have evolved photosynthesis in the first place, it was later commandeered by plants. But then plant cells are really, it seems, just some cyanobacteria that got together to live in a commune – a symbiotic relationship. Bacterial cells are basically little more than bags of DNA, and as such are called prokaryotic cells. Algal, plant, animal and fungal cells are far more complex eukaryotic cells, containing internal structures called organelles, like a nucleus, mitochondria and chloroplasts. These internal organelles are thought to have once been free-living, independent bacterial cells that migrated into a host cell to live in symbiosis, each cyanobacterium benefitting from the communal relationship. One such is the chloroplast that conducts photosynthesis.

Microbialites are made primarily by a suite of bacteria, and not all are cyanobacteria. Rather, they are an elegant cascade of different types that eventually, layer by layer, construct a microbialite.[3] The opening movement starts with filamentous cyanobacteria wrapping themselves around minute grains of sand, melding them together with sticky mucus, then standing vertically to attention. In the second movement it's time for the cyanobacteria to take a break. They are replaced by 'heterotrophic' bacteria – the organic sludge degraders. These lay a sticky sheet of mucus (extracellular polymeric substances) made from excretions and dead bacteria on top of the cyanobacteria. The third movement sees a further group of bacteria come to the party – sulphate-reducing bacteria. These feed on the sticky mucus film. Within this mucus, mineralization takes place, promoting the growth of the calcium carbonate mineral aragonite. The fourth and

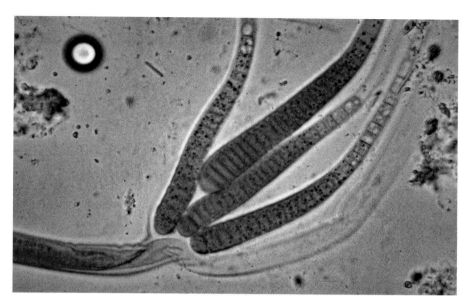

Filamentous cyanobacteria, Lake Clifton, Western Australia.

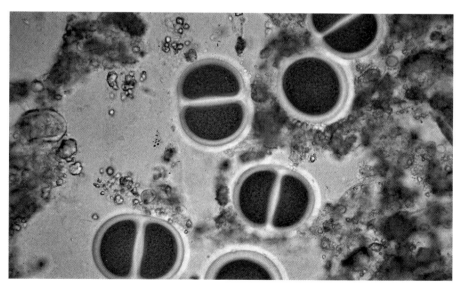

Coccoid cyanobacteria, Lake Clifton, Western Australia.

final movement sees a recapitulation, in true sonata form, of the cyanobacteria, though modified as round shell-like forms that bore through the aragonite crust, leaving behind tiny tunnels into which new aragonite can grow – a sort of bacterial reinforced concrete. Over and over again, layer by exceedingly minute layer, this bacterial symphony is replayed. All this to construct these mounds of limestone as they reach into the light, and in the process convert carbon dioxide from the atmosphere into rock, using dissolved minerals derived from the groundwater. The by-product is oxygen, our breath of life, liberated into the lake water before drifting, silently, into the atmosphere.

The great unknown in this bacterial saga has long been the actual process of mineralization. What triggers the aragonite to crystallize and form the building blocks of the stromatolitic structure in the first place? Recently a novel idea has been mooted. High-powered scanning electron microscopes have revealed tiny structures called nanospheres embedded within bacteria in the crystallizing mucus. In size they are between 20 and 300 nanometres. To put this into perspective, the hair on your head is about 70,000 nanometres wide. The numbers present are huge. There may be up to 28 billion nanospheres in each gram of gelatinous microbial mat. They seem to have faceted surfaces, rather like a football. In all likelihood, these are viruses acting as 'seed' structures around which the calcium carbonate can crystallize.[4]

* * *

BARELY A DAY GOES BY when the wind isn't blowing across Lake Thetis, either from the parched land to the east or from the vast expanse of the warm Indian Ocean in the west. Often it will blow from the east in the morning, then return the compliment from the west in the afternoon. Protein liberated from the microbial mats that form the microbialites is blown into clouds of frothy spume by these metronomic winds. The spume dances back and forth across the lake like bubbles set free from a frenetic, overflowing washing machine. Ultimately it lands on the shore. Beyond this spume-lined, rocky strand lies another, completely different ecosystem – one dominated by plants, insects, birds, mammals and fungi. It is such a highly distinctive ecosystem that it has been given its own name – *kwongan* (some spell it *kwongkan*). The word is the local Aboriginal Noongar name for 'sand plain'. Yet despite its being separated in an evolutionary sense from the microbialites of Lake Thetis by billions of years, these two ecosystems, surprisingly, have something in common: both rely on the geological vagaries of this part of the world for their existence, in particular the fact that much of Western Australia has been tectonically asleep for billions of years. Occasionally, every few hundred million years or so, sea levels have risen and flooded the land as if the nearby ancient crystalline mass of the Yilgarn Block to the east has shrugged its shoulder and caused a gentle uplift. But there have been no Alpine- or Himalayan-type mountains here for a very, very long time. Just the relentless activity of erosion, both physical and biological, wearing down the old rocks, grain by ancient grain, and all the while stripping the surface soil of almost all nutrients and trace elements.

Even though they are among the most nutrient-deprived on Earth, these 'soils' support a spectacular diversity of flowering plants (one of the richest on Earth), whose mostly endemic 8,000 or so species in the southwestern part of the state have adapted to these conditions over tens of millions of years. At the same time, it has allowed the cyanobacterial-dominated microbial ecosystem to survive. Nutrient-deficient groundwater flowing through coastal sands has allowed the cyanobacteria to thrive, instead of nutrient-dependent algae. It is this particularly nutrient-deficient environment that links the two disparate ecosystems together. Between them, one dominated by single-celled organisms, the other by a bewildering complexity of multicellular organisms, lies more than 3 billion years of evolution of a coruscating array of ecosystems – communities of organisms interacting with, responding to and even adapting their physical environment in often quite substantial ways, such as those that create rocks.

In the Beginning

About 1,300 kilometres (810 mi.) northeast of Lake Thetis, deep in the heart of the Pilbara region of Western Australia, lies an area known as the North Pole Dome – a fine example of the Australian sense of humour, as this is one of the hottest places on Earth. If the thermometer dips below 40°C (104°F), people start digging out their jumpers. Here, ragged ridges of orange, iron-stained rocks cut through the spiky spinifex grass. The rocks, at first glance, are less than inspiring. Beneath the orange veneer they are mainly just many shades of grey, from off-white to almost black. The occasional thin red layer provides some relief.

In 1977 a young geologist named John Dunlop was presented with the daunting task of trying to decipher the history of the formation of these ancient rocks. Thought to be in the region of 3.43 billion years old, they are among the oldest known rocks on Earth. The question that needed to be answered was: what are they able to reveal about the primeval Earth? To unravel this ancient story, Dunlop needed more than the words of the rock – their colour, texture and composition. He needed their geological syntax to weave them into a suitable labyrinthine tale. And this was to come from structures encased within the rocks.

The first challenge in such old geological terrains is to establish whether the rocks, as they exist today, reflect the original sediments from which they formed. Dunlop realized that this rock unit (now called the Strelley Pool Formation) comprised alternating layers of barite and chert. Barite is barium sulphate; chert, microcrystalline silica – think flint. Rocks are often palimpsests, where ghostly images have been overwritten by later geological events. Dunlop, working with colleague Roger Buick, realized that within the barite were pseudomorphs of gypsum crystals. In other words, the barium sulphate had replaced the original gypsum, but retained its crystal form.[5] Calcium sulphate had been replaced by barium sulphate. Beds of gypsum usually form, along with natural salt (sodium chloride), in evaporating shallow marine embayments, the high evaporation rates produced by hot, dry climates.

The chert has its own relict textures – patches of original carbonate, in the form of dolomite (calcium magnesium carbonate). So, the cyclical, alternating beds of barite and chert that form North Pole's jutting ridges had originally been layers of gypsum and a carbonate. Where these chemical sediments were deposited finds parallels in warm, shallow seas today, such as in the Arabian Gulf. Alternating layers of evaporites and carbonates tell of cyclical rises and falls of sea level 3.43 billion years ago, in much the same way as sea levels oscillate today. As the sea level falls, evaporites form. As it rises, shallow marine carbonate sediments are built up. But it seems that an unseen factor was helping the sediments to accumulate: the early flickerings of life.

Dunlop and Buick further refined the syntax in this oldest of geological stories as a result of a remarkable discovery. During his fieldwork Dunlop observed that the layering of the different beds of rock was anything but straightforward; often it was irregular and

Egg-carton-sized microbialites, 3.43 billion years old, Pilbara Region,
Western Australia.

wavy. But it was a small, simple domed structure about the size and
shape of a large cabbage that he found, which completed the tale.[6]
Dunlop realized he had seen one like this before. Yet it was in a completely different context – in the coastal embayment of Shark Bay
1,000 kilometres (620 mi.) away to the southwest. In fact, he had seen
more than one here. He had seen thousands. Thousands of actively
growing microbialites. Dunlop realized he had discovered not only
the oldest known microbialite in the ancient rocks of the Pilbara, but
the earliest concrete evidence for life on Earth.

With the realization that these most ancient of Precambrian
rocks in the North Pole Dome harboured microbialites, the floodgates

opened, and researchers from all over the world teemed over this part of the Pilbara region of Western Australia searching for more. Once people's eyes had been opened to the existence of microbialites in these rocks, these structures began popping out with increasing frequency. Soon, as many as seven different types other than domes had been discovered. Some were somewhat phallic-shaped conical forms; others resembled mattresses of egg cartons. A few even branched, like hands, pointing skywards to the light.[7] Even in the early history of life such a variety of complex forms had developed, suggesting that their evolutionary history must have started even before the Strelley Pool microbialites were being microbially constructed. These rocks, 3.43 billion years old, didn't mark the beginning of life. It was well under way by then. So, was the diversity expressed by these microbialites a reflection of the prolonged evolution of different suites of microbial communities, each with its own construction style? Or was it a response to subtle variations in local environmental conditions at the time – changing water depth (which the cycles of barite and chert show were occurring) or the impact of waves or currents, moulding different microbialite shapes in these primordial seas?

What they all have in common, though, is that their shapes were crafted by the activities of microbial communities, searching for light, in just the same way as their modern-day counterparts do today. This was evidence, it could be said, of construction of these small stone monoliths by photosynthetic microbes, then as now. Unfortunately, fossil microbialite, both stromatolites and thrombolites, are notorious for not preserving the microbes that make them. This is despite a great deal of effort given to searching for fossilized cells in the rocks. It may seem like a fruitless task, but such fossils are not uncommon in younger Precambrian rocks. Time and again, tiny filamentous or shell-like structures preserved in the North Pole Dome cherts have been touted as evidence for the presence of cyanobacteria in these rocks. All had been either doubted or dismissed as fantasy. Then, in 2019, researchers working on microbialites in the North Pole Dome area, in the even older Dresser Formation, made a crucial discovery.[8] Deposited about 70 million years before the Strelley Pool Formation sediments, microbialites were discovered that revealed a glimpse into a 3.5-billion-year-old microbial world. This came in the form of organic filaments, found by drilling into the exposed rock, through its outer, weathered crust into fresh microbialites.

Should you ever have the opportunity to get up close and personal with a modern microbial mat, pluck up the courage to plunge a finger a few centimetres into its gelatinous outer surface. You will be greeted by an olfactory assault the likes of which will convince you never to do so again. Here, where your finger has paused, oxygen doesn't exist. In this microbial mélange live anaerobic sulphur bacteria, with a propensity for producing the hydrogen sulphide – rotten-egg gas – that you have released. When such microbial mats lithify into rock, the sulphur often binds to iron to form the iron sulphide mineral pyrite, better known as 'fool's gold'. This is what geologist Raphael Baumgartner and his colleagues observed in layers within the Dresser Formation microbialites.[9] In places, the pyrite had preserved filaments and strands of organic matter that played a pivotal role in the formation of the microbialites. These were identical to the sheaths that wrap around filamentous bacteria in the extracellular polymeric substances – secretions from bacterial aggregations – present in modern-day microbial mats. So here, hidden in the sun-seared rocks of the North Pole Dome, the earliest evidence for life on Earth was discovered. And 1,000 kilometres (620 mi.) or so to the southwest, microbialites are doing their best to demonstrate a spectacular continuity of life. Nearly 3.5 billion years after the North

Pole microbialites began to form, they are still tirelessly growing in a warm, shallow sea. Time moves slowly in Western Australia.

Concrete Cauliflowers

In August 1699 the explorer and naturalist William Dampier, on his second visit to Western Australia, sailed into a large bay between the northwest Australian mainland and a large island (subsequently called Dirk Hartog Island). He noted: 'The Sea-fish that we saw here are chiefly Sharks. There are abundance of them in this particular Sound, that I therefore gave it the Name of *Shark's Bay*.'[10] Indeed, since Dampier's observation, 28 species of sharks have been recorded in Shark Bay, along with pods (or more accurately, apparently, 'nut-clusters') of dugongs that mooch around in its warm waters. At its southern end the water circulation is poor, due to a shallow ridge that runs east–west across the bay, restricting current flow. In this area, known as Hamelin Pool, the dugongs and sharks dare not venture. Indeed, few species do, because high evaporation of the sea in this almost-enclosed basin has raised the salinity to about twice that of normal seawater. Although a few small fish manage to survive in this shallow body of water, it is the domain of two completely different creatures: large sea snakes that like to get up close and personal if you choose to snorkel in these salty waters, and a species of tiny cockle called *Fragum erugatum*. No larger than your thumbnail, what it lacks in size the cockle more than compensates for in numbers. For the last 5,000 years or so, trillions upon trillions of these little molluscs have lived close to the shore in this salty water in shallow burrows. Around the margin of Hamelin Pool their empty shells litter the coastline. They have piled up in this cockle cemetery in accumulations metres deep, extending for many kilometres around the pool. Known as coquinas, these shell beds have formed into about 26 roughly parallel ridges, each thrown up every four hundred years or so by, it is thought, a succession of intense tropical cyclones barrelling in from the northwest.[11] Some, those furthest inland, may even have been piled up by the mega-tsunamis that occasionally sweep over the coast. Evidence for these comes from nearby sites, where huge blocks of limestone, some the size of buses, have been thrown up on top of the 15-metre-high (50 ft) cliffs.[12]

Because acid rainwater has percolated through the porous coquina over thousands of years, partial dissolution of the carbonate shells occurs during wetter times of the year, enriching the groundwater with bicarbonate ions. As the coquina dries out in the hot summers, calcium carbonate cements the shells together, forming a tough rock known as coquinite. It is strong enough to be used as an effective building stone, and has been quarried for many years. However, the bicarbonate-charged groundwater has played an even greater part in the significance of Hamelin Pool, as it is the lifeblood for the greatest accumulation of actively growing microbialites on Earth. The cockles have, effectively, been recycled as microbialites.

In the same way as Dunlop and Buick's insight bridged the divide between their 3.43-billion-year-old lump of rock and modern, growing microbialites, so too, some 25 years earlier, in the mid-1950s, Western Australian geologists made similar monumental intellectual leaps. They realized that the strange mounds (about the size of a squatting cocker spaniel) they saw emerging from the shallow water of Hamelin Pool were identical to the ancient Precambrian structures with which they were familiar in the old rocks found in inland Western Australia. Before the presence of actively forming Hamelin Pool microbialites had leapt into these geologists' consciousness, they were known only from the geological record. But there they were, slowly growing in the warm waters, twice the normal marine salinity, at the southern end of Shark Bay. Soft and gelatinous on the outside, burping out little bubbles of oxygen into the water, inside they are solid rock.

* * *

IF YOU HAVE EVER HAD THE MISFORTUNE of forgetting about a cauliflower that has been lurking unseen in your fridge for a few months, you should know that, when finally located and extracted, it will bear an uncanny resemblance to a Hamelin Pool microbialite. In an area of about 1,400 square kilometres (540 sq. mi., about the size of Greater London), Hamelin Pool's 135-kilometre-long (84 mi.) shoreline is covered almost completely by microbial mats and microbialites, many resembling the long-forgotten cauliflower. The difference between the mats and microbialites is that while the former have failed to mineralize, the latter are hard rock. Despite being found in a wide range of environments – in thermal springs, like those found in Yellowstone National Park; in lakes from Antarctica to the high Andes in Argentina; in shallow marine settings, such as Bahaman reefs; and in alkaline lakes in east Africa – there is nowhere else on Earth like Hamelin Pool. Crunching across the coquina beach to the edge of the field of microbialites is like passing back in time a few billion years, when such structures were the only evidence for life on Earth.

Microbialites have been growing in Hamelin Pool for about 2,000 years, and while it may seem that one looks just like another, research into the Hamelin Pool microbialites has revealed this to be far from the case. Closest to the shore, where the microbial mats are only covered by the sea at high tide, the mats are poorly mineralized. Whether this is because they are starved of the mineral-charged groundwater

trickling in from the shelly shore, or a paucity of appropriate mineralizing microbial cells, is not clear. These mats, smooth in places, pustular in others, are dominated by filamentous cyanobacteria.

To pass into slightly deeper water is to move into the realm of the microbialites. Those rocky mounds with gelatinous outer covers growing closest to the shore are fully covered at high tide but exposed at low tide. Unlike the mats, the microbial community that dominates these microbialites is the shell-like coccoid cyanobacteria.[13] These are the manufacturers of the reinforced-concrete-like calcium carbonate mineral aragonite, of which the microbialite is composed. Inside these mounds of rock there is no coherent structure – no layers – just a clotted mass. They are thrombolites. By contrast, in deeper water, where the microbialites are eternally submerged, they are composed of dome-shaped layers: stromatolites.

Microbialites in Hamelin Pool grow extremely slowly. Growth rates are estimated to be less than half a millimetre a year. So individual mounds, most of which are about knee-high, may be up to 1,000 years old. The deepest microbialites at Hamelin Pool, which grow in water about 3.5 metres (11½ ft) deep, reach the dizzying heights of 1 metre (3 ft). In reefs found in the Bahamas, where they grow in water of normal salinity, microbialites are even larger, up to 2 metres high, and grow in water up to 10 metres (33 ft) deep.

What could be better, from the perspective of trying to understand past ecosystems, than to have such a modern analogue – particularly when subtle environmental changes can result in noticeable morphological differences? But care must be taken when trying to make simplistic interpretations of the environment in which a rock formed on the basis of its modern counterpart. The Hamelin

Microbialites growing in hypersaline water, Hamelin Pool, Shark Bay, Western Australia.

Microbialites growing in brackish water, Lake Clifton, Western Australia. The posts are about 1 metre high.

Pool microbialites were the first modern ones to be discovered, in the 1950s. Because they form in hypersaline water that is twice the normal marine salinity, it was assumed that these were the conditions under which all microbialites grow. Such an assumption, however, was a great mistake. In Western Australia alone, actively growing microbialites have been found not only in Lake Thetis, where the water is of normal marine salinity, but in the brackish waters of Lake Clifton and in the freshwater Lake Richmond, located in the southern suburbs of Perth. Salinity, it would appear, has no impact whatsoever on the growth and development of microbialites. The critical factor in their growth is the presence of groundwater supersaturated with bicarbonate ions (usually calcium) and impoverished in nutrients such as nitrogen and phosphorus.

Rise and Fall

The North Pole Dome microbialites may only be about 1,000 kilometres from those in Hamelin Pool (a short distance in Western Australian terms), but they are separated by an unfathomably immense period of time – nearly 3.5 billion years. All the while, crumbling continents have gone on their tectonic merry-go-rounds. Oceans have risen and perished. And Earth has swung time and again between hot greenhouse and frozen icehouse worlds. Yet, through all this, microbialites of all types have marched on. For the first 3 billion years of their existence, microbialites provided almost the only evidence that life existed on this planet. This was the Age of Bacteria. It wasn't until the evolutionary diversification of animals and plants in the oceans, between 500 and 600 million years ago, that the dominance of microbialites on Earth began to fade away.

The period from 4 billion years ago to 2.5 billion, during which time these microbial structures made their first appearance, is called the Archaean (sometimes Archean) Eon. It was during this time that the Pilbara Craton was formed. One of the oldest pieces of early continental crust on Earth, it is more than 325,000 square kilometres in area (125,500 sq. mi., about the size of Germany) and consists of domes of granites that were brought to the surface about 3.7 billion years ago. This was a time before continents were the rigid plates that they are today, and most were probably very transient entities. Earth was still so hot that crustal material behaved more like a bar of chocolate that you've had in your pocket for a few hours than one straight out of the fridge. Today, granites, which have a lower density than basalts, occur deep in the crust, while basalts form on the surface. During the Archaean, however, this was often the other way round, as the heavier basalts easily sank beneath the granites, which were forced to the surface. Hence the formation of domes like the North Pole Dome. On and around these early granitic land masses, life gained a foothold in shallow seas and even in lakes that insinuated themselves across the land surface.[14] Into these aqueous areas sediments accumulated, having been shed from the nascent lands, and in warm, quiet bodies of water limestones began to form, ably assisted by the recently evolved microbial communities.

By the Late Archaean, about 2.7 billion years ago, the Pilbara Craton was an emergent land mass upon which volcanoes spewed out vast amounts of lava. Following its consolidation into rock, physical weathering by the action of wind and water produced sands, silts and gravels that were washed into the shallow seas around the craton or accumulated in depressions on the land. Evidence for this process comes from a unit of rocks called the Tumbiana Formation. Up to 320 metres (1,050 ft) thick, these layers of basalt, sandstone, siltstone and occasional limestone extend for more than 680 kilometres (422 mi.) on the craton. One particular 30- to 50-metre-thick (100–165 ft) sequence called the Meentheena Member consists of alternating beds of silica-rich sediments, such as sandstones and siltstones, with limestones. These rocks represent one of the world's earliest documented lake systems. At more than 680 kilometres (423 mi.) long, this sequence of lakes was longer even than the second-longest lake on Earth today, Lake Tanganyika. And the limestones that formed in the shallow water along the lake's margins reveal one of the most extensive occurrences of microbialites in the geological record.

The discovery of and documentation and research into ancient Archaean microbialites over the last two decades has revealed that they steadily increased in size and diversity through this period. They created not only gently rippled layers, but domes, nodules, columns (some that branch), cones and spheroidal forms called oncolites. All these occur within the Tumbiana Formation limestones, where more than ten distinct types have been recognized.[15] In places, they crowd close together and merge to form extensive reefs up to 16 kilometres (10 mi.) long along the margins of the lakes. Layers of silica-rich sands and silty sediments alternate with limestones. These tell the story of rivers running into lakes: sometimes flowing vigorously and bringing the siliceous sediments; at other times drying up, allowing the formation of microbialites. Like today, there must have been cycles of abundant rainfall punctuated by periods of drought. How long each part of the cycle persisted remains unknown. No animals or plants lived in the Tumbiana lake at this time, or on the bare, windswept land that surrounded it. The only signs of life were the communities of microbes, steadfastly photosynthesizing and, millimetre by millimetre, creating rocks resilient enough to last for 2.7 billion years – rocks that, like those in the North Pole Dome, were constructed by microbes in response to the sunlight falling onto these ancient lakes.

By the end of the Archaean, 2.5 billion years ago, there was a worldwide expansion in large, stable shallow-water environments, greatly increasing the potential area for growth of microbialites.

Eucapsiphora leakensis, a 1.64-billion-year-old branching columnar microbialite growing from microbial mat in the Pilbara region, Western Australia.

For the following billion years or so, during the early part of the Proterozoic Eon (2.5–0.54 billion years ago), they increased greatly in diversity, reaching the pinnacle of their development, in terms of abundance, diversity and size, between about 1.5 and 1 billion years ago. Some stromatolites were columns just a few millimetres high; others were bigger than a football field. This increase in diversity probably reflects subtle adaptations to different environmental conditions. For instance, columnar forms are thought to have grown in lagoons, while broad domes with internal branching, very much like the modern ones in Lake Thetis, occupied even shallower water. By 1 billion years ago more than 350 types of stromatolite are known to have existed worldwide. This diversification is likely a response to a large increase in bacterial diversity, particularly of cyanobacteria.

In addition to looking at the great sweep of evolutionary change in microbialites over billions of years, it is also possible to get up close and personal with individuals of some of these ancient colonies and see how they grew from their first ripples on a microbial mat to a towering layered stromatolitic structure. When a bed of stromatolites is followed along a single horizon, there is often little change in its appearance. Presumably conditions suitable for a particular type of growth were much the same in one place as in another. However, a consistent theme over billions of years is that each stromatolitic structure frequently changed its appearance substantially as it developed. From a basal, near-horizontal layer of microbial mat, domed structures developed, as the photosynthetic cyanobacteria that construct the layered stromatolites were attracted to the light. Moving further upwards, finger-like columns pointing skywards emerged. Finally, these branched to the outer surface of the stromatolite, still seeking the light. There has never been a satisfactory explanation for why this sequence has occurred repeatedly, over vast tracts of time, all over the world. One possibility is that the final development of branching columns would greatly increase the surface area of the structure, providing potential for a larger number of microbes on the surface of each stromatolitic microbialite.

Following their rise in diversity and abundance up to about 1 billion years ago, microbialites went into steep 500-million-year decline – in response, it has been argued, to the evolutionary diversification of animals. Animals came and microbialites went. Why such a mutual exclusivity? One much-touted theory is that early animals grazed on the microbial mats and microbialite surfaces, effectively eating them out of house and home. This idea is based on the absence of animals grazing on the Hamelin Pool microbialites, due to the hypersaline environment in which they grew, but in which grazers do not live.

However, as I have pointed out, there is a major problem with this hypothesis as microbialites will also happily grow in lakes of normal salinity, brackish-water lakes, even in fresh water.[16] Importantly, it has been shown that in Lake Clifton – a brackish-water lake 100 kilometres (62 mi.) south of Perth in Western Australia, where vast swathes of thrombolitic microbialites have created an extensive 8-kilometre-long (5 mi.) reef fringing the lake – grazers, mainly crustaceans (isopods, amphipods and ostracods) and worms (polychaetes and nematodes), munch on the microbialites with no detrimental effect.

A more probable explanation for the sharp decline in the abundance of microbialites at the end of the Precambrian is a fundamental change in the nutrient status of the water in which they grew. At this time, plankton in the ocean proliferated, and with them levels of nitrogen and phosphorus rose substantially. Active ocean circulation brought this increase in nutrient level onto shallow marine shelves, where microbialites formed.[17] Today it is well known that these structures will only flourish in nutrient-poor environments. Increasing nutrient levels consequently had a negative effect on the microbial communities that construct them.

This has been shown only too well in Lake Clifton. The increase in human population density close to the lake, combined with nearby agricultural practices, has caused microbialite growth to severely diminish. Rising phosphate levels in groundwater feeding the lake during the last few decades have stimulated excessive growth of the green alga *Cladophora*, including on the submerged microbialites themselves. This has effectively disrupted cyanobacterial growth and, therefore, formation of the microbial structures. The long-term effect is likely to be the death of the entire microbial community that constructs the microbialites. Sadly, this could be, on a very small scale, a repeat performance of the environmental changes that impacted microbialite diversity at the end of the Precambrian – a shift that led to their relegation from a once-dominant feature of the geological landscape to relative rarity in the following 500 million years.

Restricted to low-nutrient environments, such as reef systems, microbialites persisted in extensive reefs that formed in Devonian times about 370 million years ago in North America, North Africa and Western Australia. They even occur in much younger rocks, such as the mere 145-million-year-old Purbeck Limestone in Dorset. In the 1820s, when our knowledge of geology was in its infancy, William Buckland (1784–1856), professor of geology at the University of Oxford and dean of Westminster, accompanied by his friend Henry De la Beche (1779–1855), was undertaking the first serious geological study of the rocks of the Dorset coastline. Both knew the area well – De la Beche lived in Lyme Regis, while Buckland had been born in nearby Axminster. Buckland was renowned as a somewhat eccentric character – that is, if you consider his hobby of attempting to

Layer of microbialites in 145-million-year-old Purbeck Limestone,
Lulworth Cove, Dorset.

eat his way through the entire animal kingdom a sign of eccentricity (Buckland considered bluebottle flies and moles the most disgusting of the many animals he ate). When offered the chance to view what was said to have been part of the petrified heart of King Louis XIV of France at a dinner at the home of the Archbishop of York, Edward Venables-Vernon-Harcourt, Buckland couldn't resist the opportunity to add it to his gastronomic list. He is reported to have observed, prior to popping it into his mouth, 'I have eaten many strange things, but I have never eaten the heart of a king before.' Despite this arguable lapse in gastronomic judgement, Buckland was a great geologist. He was the first in England to embrace Louis Agassiz's ideas on the role of glaciers in moulding the landscape, and his seminal work on cave fauna in Yorkshire launched the science of palaeoecology. Although the Dorset coast might seem somewhat more prosaic in character, what he and De la Beche found there was, nevertheless, intriguing.

Lulworth Cove, an almost-enclosed embayment resembling a bite taken out of the coastline, would have looked much the same when Buckland and De la Beche visited it as it does today. Climbing out of the cove on its eastern side and up onto a resistant limestone ridge that parallels the coast, their only danger would have been the risk of falling off the cliff into the sea. Today there is an added risk. It is a British Army shooting range. Fortunately, the army rests at weekends, and it is possible to gain access to what Buckland and De la Beche described as a fossil forest, clinging to the cliffs.

Tracing the same beds from the Isle of Portland to the west, they documented layers of limestone interspersed with a prominent fossil soil horizon (known as the Dirt Bed), in which they found the silicified remains of coniferous tree trunks and cycads. Above this they describe a thin limestone band, calling it the 'Soft Cap',[18] within which are microbialites, both stromatolites and thrombolites, that had once grown in a brackish-water lake.[19] As you descend the cliffs on steps now conveniently located on the cliff's edge, it is possible to get a spectacular drone's-eye view of some of these rocks that form domes a metre or so across in the Purbeck Limestone. Some of the microbialites have a central depression, and these have been explained since Buckland and De la Beche's day as the site where trees once stood, the microbialite growing as a ring around the tree. While in lower layers silicified trees have indeed been found, these tyre-like microbial mounds of rock are much the same as those that encircle Lake Thetis back in Western Australia. The only difference is that while seabirds now perch on the Lake Thetis microbialites, 145 million years ago the same structures in the Purbeck Limestone would have more likely been perching places for flocks of pterosaurs.

While cyanobacterial-dominated microbial communities have relentlessly produced limestone for more than 3 billion years, once other groups of organisms evolved – fungi, plants and animals – they also got in on the act, playing significant roles in limestone formation. Foremost among these were invertebrate animals, without which so many types of limestone would just never have existed.

THREE

ROCK OF AGES

In recent years, carbon dioxide has had something of a bad press. A main contributor to climate change, 'that dangerous gas' is said to induce a range of effects, from extreme weather to food supply disruptions, increased wildfires and ocean acidification. The 'carbon dioxide problem' list goes on. You could, perhaps, be forgiven for thinking that the world might just be a better place if we got rid of the stuff altogether, or if it had never existed on Earth in the first place. I mean, just what good did carbon dioxide ever do for us? Well …

On early Earth, when the Sun was weak, it helped stop the planet from becoming an enormous cosmic snowball. It facilitated photosynthesis, without which animals and plants would never have formed. It enabled a cornucopia of invertebrates to biomineralize and produce hard, protective shells. Imagine being unable to tuck into a plate of oysters. Or sumptuous lobsters. Indeed, imagine a world in which the oceans had very few marine invertebrates at all. For without such animals, many limestones would never have formed. Some, like chalk, rely on algae for their formation. Microbialites, on the other hand, are dependent on the activity of suites of microbes, while other limestones are the mashed-up remains of the shells of biomineralizing invertebrate animals. Others come ready-made as rocks, like corals, that secrete calcium carbonate. All draw their carbonate from carbon dioxide dissolved in water. Oceans and lakes are sinks for carbon dioxide, some of which becomes sequestered into rocks, thanks to the activity of so many types of organisms. As such, limestone is a reflection of the evolution of life, from its first tentative beginnings to today. A true rock of ages. Maybe there's something to be said for carbon dioxide after all.

M ost societies, since the dawn of humanity, have embraced the concept of a supernatural world of the dead – the underworld. It is the underworld to where souls of those who have passed on go. But when the minds of geologists in the eighteenth century began to strip the land bare and delve into the rocks beneath the veneer of meadows and forests, they exposed a real underworld. This one often contained the remains of the dead – rocks teeming with the petrified remains of organisms that had once been living, respiring and reproducing organisms. Of all rock types, limestone embraced this enthusiastically, being either composed of the remains of the dead or created by the activity of organisms that are now long dead. But to lump all these rocks together under 'limestone' is a disservice to the amazing variety of rocks that labour under this one name. The spell-check on my computer gets very upset when I type the word 'limestones'. A squiggly red line appears telling me that there is no such plural. It thinks there is only one kind of limestone. The computer could not be more wrong . . .

First Limestones

Carbon dioxide is repurposed into limestone by the activity of virtually every major group of organisms, ranging from viruses to bacteria, fungi, algae and animals. As such, the nature of limestones is intimately bound up with the evolution of life. They are a product of their time. Limestone is made of either calcium carbonate or a mixture of calcium and magnesium carbonate. If calcium and magnesium each comprise about 50 per cent then the rock is called dolomite, or dolostone. Limestone and dolomite are often collectively referred to as 'carbonate rocks'. They have a history that extends through more than 3.5 billion years, spanning the Precambrian to the present day.

From the time when continents first began to accrete on Earth to the first glimmerings of biomineralizing metazoan life, about 600 million years ago, limestones were being formed in the oceans. Within these rocks are chemical, biological and physical clues to environmental change, ancient tectonic regimes and the evolution of life.[1] Their physical attributes provide insights into tectonic subsidence and sea-level fluctuations; their chemical make-up indicates the rate of burial of carbon extracted from the atmosphere and the growth of continents; and the fossils they contain tell the story of the evolution of microbial communities and ecosystems.

These microbial communities were dominated by bacteria in the Archaean Eon (4–2.5 billion years ago), and continued to diversify during the ensuing Proterozoic Eon, up to a little over half a billion years ago. They were the dominant group of organisms that constructed limestone structures in the form of microbialites. Indeed, until algae came on the scene, they were the only organisms thriving in these ancient oceans. Yet many of the first limestones may not have had an organic component to them at all. Encrustations of the two polymorphs of calcium carbonate, calcite and aragonite, grew on the Archaean and early Proterozoic seafloors in open marine conditions. They precipitated as divergent crystal fans, with radii in the order of tens to even hundreds of centimetres. These crystalline limestones formed beds many metres thick that can be traced over 100 kilometres (62 mi.).[2] This type of limestone steadily decreased in frequency through the Proterozoic. The reason for the transition from abiotic to largely biotically formed limestones (in the form of microbialites) through the middle to late Precambrian is likely connected with the transition from an oxygen-free to an oxygenated atmosphere. During much of the Archaean, the atmosphere was primarily composed of nitrogen, methane and carbon dioxide. As a consequence, the seas were oversaturated with calcium carbonate, resulting in a far more alkaline ocean than today that facilitated direct precipitation on the seafloor. As the partial pressure of oxygen increased and carbon dioxide decreased during the later Precambrian, so the transition from abiotic to biotically produced calcium carbonate took place. Calcium carbonate precipitation was also inhibited by the presence of ferrous iron in the early Proterozoic seas. The reduction in this more soluble form of iron in the oceans, leading to the deposition of banded ironstones, led to a decline in calcium carbonate precipitation on the ocean floor.

Through the Proterozoic (in other words from 2.5 billion to 540 million years ago) limestones progressively came to be constructed from microbialites. Many expressed themselves in their capacity to construct large reef structures: in their overall architecture, limestones show similarities to modern day coral-dominated reefs, forming a range of structures from major barrier reefs to pinnacle reefs and patch reefs.[3] These reefs were able to grow from deep, quiet water settings to those in shallow water, where wave action was more intense. Physical abrasion of the microbialite reefs in such an environment generated carbonate sediment. Most of these reefs were formed of layered, stromatolitic microbialites that grow by both the trapping and binding of sediment and the precipitation of calcium carbonate. The extensive structures were formed by the coalescence of domal stromatolites as they increased in size.

Later in the Proterozoic the other major type of microbialite, thrombolites, came to play an increasing role in microbialite reef development. Growing in shallow, sub-tidal marine environments, these more irregular 'clotted' structures contain a greater porosity compared with stromatolites. These pores enable growth of calcite cements as well as sites for the accumulation of fine carbonate sediment. An additional factor in the development of more porous reefs was the evolution of what are termed 'calcimicrobes', a range of different microbial elements capable of inducing the precipitation of calcium carbonate. These were to become important in the development of reefs hundreds of millions of years later in Palaeozoic times. The appearance of thrombolites and calcimicrobes about 1 billion years ago corresponded with a decline in diversity of stromatolites. It may not have been a coincidence, but with the appearance of these more porous reefs, the first calcified metazoans appeared in the form of simple, small calcium carbonate tubes called *Cloudina*. These are found within the clotted thrombolites as well as between individual domes. Both whole and broken shells began to contribute to carbonate sediments.

Ovary of a Herring

As well as these large framework microbialite carbonates, shallow Precambrian seas, even from early Archaean times, were awash with a type of carbonate sediment that has persisted for more than 3 billion years. Cemented by calcium carbonate derived from water percolating through the pore space in this sediment, the result has been the formation of a rock that has been used to construct cathedrals and palaces: oolitic limestone.

When presented with a piece of such limestone from Nunnington in Yorkshire for his collection, John Woodward described it as 'a mass of Stone made up almost entirely of the little round Pellicules of the Ova of Fishes, filled with a fine stoney Matter'.[4] Woodward's interpretation was not unreasonable, for oolitic limestone bears a more than passing resemblance to petrified caviar, and sometimes it is just as black. Formation of ooids has long been the subject of much

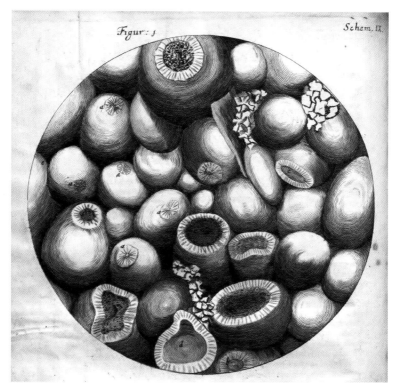

Illustration of 160-million-year-old Middle Jurassic Ketton limestone oolites by Robert Hooke, who thought they looked like the 'Ovary of a Herring'. Published in his book *Micrographia* (1665).

discussion. The first mind turned to this rather strange rock was curator of experiments at the Royal Society in the seventeenth century and professor of geometry at Gresham College in London, the great polymath Robert Hooke.[5] Using the newly created microscope, Hooke peered closely at a small piece of oolitic limestone, which he described in his *Micrographia* (1665) as having been 'digg'd out of a Quarry' and 'brought from Kettering, in Northampton-shire'. It is now thought that his specimen actually came from a quarry at Ketton in Rutland. The Ketton Quarry is of Jurassic age and is a particularly pure oolite, consisting of little more than a mass of spheres that vary in width from about 0.5 to 1 millimetre. Ideal as a building stone, it forms the Lincolnshire Limestone Formation and has been used in the construction of many Cambridge colleges, notably Downing College and the Wren Library in Trinity College. Hooke was much attracted to this rock as an ideal candidate for perusal down his microscope, as it 'has a grain altogether admirable'.

> It is made up of an innumerable company of small bodies, not all of the same cize [*sic*] or shape, but for the most part, not much differing from a Globular form, nor exceed they one another in Diameter above three or four times; they appear to the eye, like the Cobb or Ovary of a *Herring*, or some smaller fishes, but for the most part, the particles seem somewhat less, and not so uniform . . . [6]

But where did it form? Hooke was quite aware that the spheres were not fish eggs. Perhaps, he thought, they had been formed in the sea. 'The object, through the *Microscope*, appears like . . . a heap of Pibbles, such as I have often seen cast up on the shore, by the working of the Sea after a great storm.' Travel to many a Caribbean island and indeed the beaches are often composed of sands made largely of little ooids. But how did they form? Hooke reasoned that it could have been

> by the working and tumblings of the Sea to and fro . . . jumbled and comminuted into such Globules as may after-wards be hardned into Flints, the lying of which one upon another, when in the Sea, being not very hard, by reason

of the weight of the incompassing fluid, may cause the undermost to be a little, though not too much, varied from a globular figure.[7]

Given that one of the present-day favoured explanations for the formation of ooids is that they form by the accretion of calcium carbonate around a minute shell or fragment of a shell, as they roll back and forth in shallow warm sea, Hooke's explanation is, in many respects, extremely prescient. Looking down his microscope, Hooke was also intrigued by what occupied the spaces between the spheres. Sometimes there were crystals of calcite, gluing the ooids together; often there appeared to be nothing but a void. Hooke suspected this meant that the rock was porous enough to allow the passage

Oolitic limestone from the Carmel Formation (Middle Jurassic) of southern Utah; mainly ooids plus a single stem ossicle from an isocrinid crinoid.

of fluids. To test this idea, he 'took a pretty large piece of this stone' and sealed it with a cement, 'save only at two opposite parts'. He then proceeded to cover one surface with his spittle. Placing his mouth against this wet side he blew hard and produced on the opposite surface an 'abundance of bubbles, which argued these pores to be open and pervious through the whole stone'.[8] Hooke's proof that the oolite was porous justified its use as a building stone. Hooke and Christopher Wren worked closely together in the rebuilding of London after the fire of 1667. One of the major stones used, especially in St Paul's Cathedral, was oolite, though from Portland in Dorset rather than from Ketton.

While such oolitic limestones are a common feature of many Jurassic rocks, in particular in northern Europe, their occurrence in the rock records extends way back to some of the earliest Archaean limestone. In the 2.7-billion-year-old lacustrine Tumbiana Formation, found in the Pilbara region of Western Australia, oolites occur interbedded with microbialites. Like many of their Jurassic counterparts, individual ooids are made of calcite and are of a similar size, barely reaching 1 millimetre in diameter. Recent studies have revealed that rather than being simply a product of the physical growth of layers of calcite as the spheres rolled around in agitated shallow water, organisms played a role in ooid formation. Analysis of the ooids has detected evidence for the existence of microbial biofilms layered within the spheres. This suggests that ooid formation 2.7 billion years ago was, like microbialites, mediated by the activity of cyanobacteria and sulphate-reducing bacteria.[9] A similar conclusion has been reached for modern-day ooids. Experiments on ooids forming in shallow-water environments around the Bahamas have revealed a complex association of their calcium carbonate structure with the products of microbial communities. Crucial in ooid formation are extracellular polymeric substances (EPS), produced by biofilm-forming bacteria. Calcium carbonate precipitates within the EPS as well as on bacterial cells within the ooids.[10] So, for at least 2.7 billion years to the present day, microbial communities have played a key role in the formation of ooids, and consequently oolitic limestones. But the dominant role of microbial communities in limestone formation for nearly 3 billion years was to change, and change relatively rapidly from a geological perspective, with the evolution of multicellular animals.

An Explosion of Life

The geological time period known as the Phanerozoic spans from the end of the Precambrian, 540 million years ago, until today, and its first 55 million years, the Cambrian Period, mark one of the most important steps in the evolution of life on Earth.[11] Early in this period, all of the major modern animal phyla either appeared or began to diversify greatly, among them groups like molluscs, echinoderms, arthropods and even our own, the chordates. Importantly, for the nature of the sedimentary record, the Cambrian is characterized by the appearance in the fossil record of a panoply of organisms that had evolved a really strange characteristic: the ability to biomineralize – in other words, to construct hard parts inside or outside their bodies. This they did by mineralizing shells, spicules and scales. While the fossil record indicates that a few groups, including sponges and the calcareous conical shell *Cloudina*, had evolved this ability during the later Proterozoic, most groups of phyla that still exist today evolved during a period of as little as 10 million years in the early Cambrian. These include many with mineralized parts, including arthropods, molluscs, the armoured slug-like halkieriids, brachiopods and echinoderms. Their remains were all to become grist to the sedimentary mill.

In rocks older than the Cambrian, those few faunas of body fossils and traces left by itinerant invertebrates moving across sandy substrates are characterized by their low diversity. But within the earliest Cambrian, which, after all, is characterized by the appearance of these new animal groups, assemblages are far more diverse. These earliest beds contain fossils of what has been termed, with great precision, the 'small shelly fauna'. This quickly gave way to an even more diverse fauna dominated by arthropods, notably trilobites, with exoskeletons made of calcium carbonate.

The early Cambrian biomineralization of many metazoans was usually based on calcium, though some groups had metabolic processes that yielded calcium phosphate, rather than calcium carbonate. Other groups, such as some sponges, experimented with other chemicals, giving up calcium altogether and generating hard parts made of silica. These, however, were not very common in the Cambrian seas. Biomineralized structures are likely to have evolved for protection in response to the evolution of predators. The top-line predator in the Cambrian was *Anomalocaris*, a near-metre-long arthropod. Evidence for predation comes from studies of Cambrian trilobites which show that many of the healed scars on trilobites' bodies match the size and shape of the mouth parts of *Anomalocaris*. Surprisingly, the trilobite predators exhibited strong right–left behavioural asymmetry, as predation scars reveal that 70 per cent were preferentially attacked on the right side. In living animals such behaviour occurs as a result of lateralized nervous systems, evidence that these systems were present even in early Cambrian arthropods.

Exactly what caused the 'Cambrian explosion' in marine life remains enigmatic. Significant changes were occurring in ocean chemistry and circulation at the time. In particular, the period in Earth history around 550 million years ago is characterized by extensive production of phosphorus in the oceans, an element crucial to meta-zoan metabolism. This rise in the level of phosphorus is likely to have occurred due to changes in global tectonics, ocean circulation and nutrient supply. Some researchers have suggested that increases in oxygen levels may, in some way, have also contributed to the evolutionary explosion of metazoan diversity. However, it is unclear to what extent these changes in ocean chemistry either affected, or were stimulated by, the evolutionary radiations, or whether they merely reflected the changing composition of the fauna and its diversity.

There is also a possibility that, rather than having abiotic causes, some of the chemical changes in the oceans may have occurred as a result of the 'Cambrian explosion' itself, arising from the evolution of more complex food webs.[12] Even by Cambrian times, marine invertebrate communities had developed a wide range of feeding strategies analogous to those in present-day marine ecosystems.[13] Moreover, marine primary productivity increased significantly in the Cambrian, leading to a major diversification of phytoplankton that are low on the food chain.[14] It may even have been driven by an increase in the diversity of zooplankton, such as small filter-feeding arthropods – a potentially critical innovation that led to the evolution of larger animals.[15] In addition to providing a food source for larger predators, zooplankton may have significantly affected the ecology of the ocean floor. Their faecal pellets are likely to have played a role in oxygenating the seafloor and increasing the availability of nutrient-rich organic compounds to organisms living on and within it.[16] Whatever the cause, the effect of the 'Cambrian explosion' on the nature of limestones was immense.

* * *

AT THE NORTHEASTERN END of the Dead Sea, tucked away in the southernmost part of Jordan, is an outcrop of early Cambrian limestone. Deposited as calcium carbonate-rich sediment roughly 520 million years ago, this limestone, little more than 12 metres (40 ft) in thickness, displays the great increase in diversity that occurred in carbonate sediments shortly after the Cambrian explosion of marine life.[17] Microbialites are still there, but only as infrequent thin layers, interspersed with fine carbonate mud. So too the oolites; some beds are still composed almost exclusively of these little millimetre-sized spheres. In other layers, however, they are replaced by the Phanerozoic newcomers – the skeletal remains of a host of recently evolved marine invertebrates, all of which secreted calcium carbonate within or around their bodies. Trilobites, primitive echinoderms, brachiopods, the now-extinct conical-shelled hyolithids and needle-shaped sponge spicules, often make up more than half of the composition of the limestone beds.

Many of the tiny shell fragments show signs of having been bored by a range of micro-organisms that inhabited the shallow marine waters, from fungi to algae and bacteria. Such activity would have generated the production of very fine carbonate mud and hastened the breakdown of these bioclasts. The sizes of the shell fragments, along with the percentage of mud within each layer of the limestone outcrop, provide pointers to the energy of the environment in which the different sediment types were formed. Beds containing a mix of ooids and shell fragments, often well sorted into specific size fractions, indicate deposition in water where there was moderate current activity, such as shallow-water sand bars. Nearshore, where there were strong currents, coarser sediments with relatively large bioclasts called rudstones would form.

Many layers also contain the signatures of these new invertebrate animals in the form of small, ovoid mud pellets called peloids. All animals, from ostriches to nematode worms, share certain, basic traits – a mouth, a gut and an anus. Food goes in at one end, is processed, and waste products are excreted at the other end. During Cambrian times a host of invertebrates, many lacking hard parts, such as priapulid and polychaete worms, evolved. Some were deposit feeders – processing the fine sediment created by bioeroding organisms and extracting nutrients from the organic component of the sediment and voiding the remainder as peloids of compressed carbonate mud. The presence of peloid-rich layers at particular horizons implies periods of high nutrient availability for the organisms processing the sediment. The more nutrients, the greater the density of the infauna and the more peloids produced.

Subtle differences in the texture of carbonate sediments, such as the amount of coarse and fine material, usually mud and ooids or shell fragments, reflects subtle variation in the environment of deposition. The resultant limestones are classified accordingly. There are mudstones, where, not surprisingly, most of the sediment comprises mud. Introduce a few shell fragments into the mud mix and the resultant limestone is called a wackestone. When the proportion of invertebrate skeletal remains greatly exceeds the mud component, the limestone that ultimately develops from the sediment is called packstone (well-packed fragments). The mud forms a supporting matrix. When the limestone is composed entirely of shell grains and there is no mud the rock is a grainstone.

These names might seem to be somewhat esoteric, but as descriptive terms for subtly different types of limestone they are valuable in being reflections of the environmental conditions under which the carbonate sediments were deposited. Early Cambrian limestones

A bioclastic limestone formed of ammonites in carbonate mud. Called 'Marston Marble', it derives from the 195-million-year-old Early Jurassic Charmouth Mudstone Formation in Somerset. This specimen was collected in the mid-19th century and became part of the private collection of James Tennant. It is now housed in the Western Australian Museum, Perth.

in Jordan, like many such sequences, tell a story about changing sea levels over half a billion years ago. The shallower the water, the higher the wave and current energy, and the more it is able to move large sediment particles, but also to winnow out the fine mud. These are transported and accumulate in deeper water. So a sequence of mudstone to wackestone to packstone to grainstone could well reflect an increase in energy of the currents as sea levels fall and the water column at that location becomes steadily shallower. The reverse sequence will develop as sea levels rise and the water deepens. Repeated cycles of these changes are common in the rock record, reflecting cyclical changes driven by regular Milanković perturbations on the scale of tens to hundreds of thousands of years. Here, in an ancient dead sea, are cycles of shelly limestones, the product of life upon the Earth and astronomical forces beyond it.

Carbonate Factories

By the end of the Cambrian Period, limestones had begun to undergo another fundamental change, as life in the oceans, upon which they owed their existence, started to transform. From about 488 to 443 million years ago, in the geological time period known as the Ordovician, three sequential, major evolutionary episodes took place in the oceans. These were to have dramatic effects on the nature of the source material for carbonate sediments, changing them forever. This was the time of what has been dubbed the 'Great Ordovician Biodiversification Event'.[18]

The first evolutionary expansion in this event took place in the tiny plankton, both zooplankton (especially graptolites, radiolarians and chitinozoans) and phytoplankton (notably a group of enigmatic organisms called acritarchs) that inhabited the oceans, from the shallowest water to the deepest. These planktonic groups continued to increase in diversity throughout much of the Ordovician. Their expansion may have been triggered by increases in inorganic nutrients taking place in the oceans, caused by the very active volcanism that was occurring at the time. This was driven by extensive movement of tectonic plates, which had the effect of generating very large quantities of CO_2 into the atmosphere. Consequently, the atmospheric level of the CO_2 was up to fifteen times higher than today, the highest it has been in the last half a billion years. Such a level, though, was critical for the maintenance of a climate favourable to life, because solar luminosity at this time was much lower than today.[19] With the expansion of plankton came a concomitant expansion of actively swimming marine invertebrates that possessed calcium carbonate shells feeding on this expanded food source – shelled nautiloid cephalopods. Their dead shells would contribute to the build-up of carbonate-rich sediment.

The second evolutionary expansion during the Great Ordovician Biodiversification Event was in those marine invertebrates that lived on and within the seafloor, whose skeletal remains contributed an increasingly major input to carbonate sediments and hence to the composition of limestones. These include immobile groups, like brachiopods and bivalves, and mobile groups such as echinoids (sea urchins), asteroids (starfish), ophiuroids (brittle stars) and gastropods.[20] The other organisms to undergo major diversifications, and which were to change the nature of both carbonate sediments and limestones dramatically, were colonial animals, especially corals and bryozoans, along with calcareous sponges called stromatoporoids. With their expansion came a transformation in the nature of reefs, from microbial to being dominated by these newly evolved colonial marine invertebrates. Reef structures, or bioherms, were quite small

The 375-million-year-old Late Devonian Great Barrier Reef, Windjana Gorge, Kimberley Region, Western Australia. Inclined beds on left, fore-reef sediments; horizonal beds on right, shallow-water back-reef. Main reef at junction of the two.

at this time, often measuring just tens of metres. But over the next 100 million years they were to increase in size and complexity to such an extent that by Late Devonian times, around 370 million years ago, the metazoan reef builders had built stupendous structures many hundreds of kilometres long.

Many of the organisms that initially diversified during the Ordovician Period may have played roles, both directly and indirectly, in drawing down CO_2 from the atmosphere and trapping it either in the carbonate sediments that they created, or in the bound reef structures that were formed: ready-made beds of limestone. Whether this reduction in greenhouse gases contributed to the reduction in global temperatures that led to a major global glaciation at the end of the Ordovician Period, as some have suggested, is debatable. Either way, the character of the limestones changed irrevocably.

* * *

IT IS LATE AFTERNOON, and the setting sun has one last task before disappearing for the night. Ahead lies a massive limestone cliff at the southern end of the Kimberley Plateau in Western Australia. It slides away from north to south as far as the eye can see. There can be little doubt that the cliff is made of limestone. Its face is scoured deeply with vertical gouges made by chemical dissolution of the rock as water has drained down the 100-metre-high (330 ft) cliff face for an infinite number of years. The setting Sun transforms the usually dull, grey cliff into a towering wall of silver and gold. One part, however, lies in shadow.

Scoring the face is a gorge, barely 200 metres (654 ft) wide; somehow it has carved its way through this massive range of limestone. How it formed in such a hard rock is a mystery. One thought is that it developed 300 million years ago, when the area, like all Gondwanan lands of the time, was covered by ice more than a kilometre thick. Beneath this, as the ice sheet slowly melted at the end of the great Palaeozoic ice age during the early Permian, meltwater gouged out the gorge. However, and whenever, it formed, this exposure of the reef underworld has offered insight into this limestone range, revealing it to be part of a huge reef system that formed for about 20 million years, between 382 and 362 million years ago.[21] Originally flanking much of the southern and western margins of the Kimberley land mass, its remains now extend for about 350 kilometres (217 mi.) along its southwestern flank. In places it is up to 50 kilometres (30 mi.) wide, and all, essentially, constructed by a menagerie of organisms from bacteria to animals.

A new day, and the long climb to the top of the range, up the side of the gorge and over jagged limestone pinnacles, is rewarded with a view of arguably the finest exposure of a fossil reef in the world. As hackneyed as it sounds, it really feels like being transported to a Late Devonian barrier reef, because staring back across the gorge are limestones that expose the structure of this ancient reef. To the right, where the Sun is just rising above the ancient Kimberley land mass, stretch horizontally bedded platform limestones that formed in the quiet, shallow water between the reef front and the land. Straight ahead is the reef front itself, barely 10 metres (33 ft) wide. While the seas that battered this ancient reef front have long gone, the limestones to the left plunge steeply to the west into what once was deep water. These formed from sediments eroded from the reef.

Each of these three areas – back-reef, reef and fore-reef – have their direct parallels in today's Great Barrier Reef that flanks the north Queensland coast. Its Devonian equivalent existed for about 15 million years. Changing locations of the reef limestones, dated using changing fossil faunas, show how the reef system periodically migrated inland as the sea level rose, then retreated as it fell. Unlike its modern equivalent, the reef was not solely constructed of coral, though it did contribute a minor part of the ancient reef system in the form of small coral heads, up to a metre in diameter, growing in the quiet back-reef waters.

Despite the rise in dominance of invertebrate marine animals during the Palaeozoic, the reef was still the domain of microbes. The dominant form was a structure called *Renalcis*, which precipitated a microbial mud of calcium carbonate. It generated a ready-made carbonate rock that formed the spine of the reef. Minor amounts of calcareous sponge-like organisms called stromatoporoids, along with true sponges, also helped formed the reef structure. The quiet back-reef limestones show an environment of fine carbonate mud, flat, platey stromatoporoids up to 5 metres (16 ft) across and corals.[22] The limestones that represent the fore-reef environment are very different from the reef and back-reef limestones. The sediment that formed them was eroded from the reef front, where there was physical destruction by wave action, as well as bioerosion by fishes and a range of invertebrates. The limestones contain a more diverse fossil fauna – brachiopods, trilobites, gastropods, bryozoans as well as stromatoporoids and deep-water stromatolites – their fragmented remains all contributing to the formation of the limestones. Periodic earthquakes dislodged enormous blocks of reef, which tumbled into the deep, dark waters and became incorporated into the fore-reef limestone. Some such blocks were up to 200 metres (218 yd) across.

Set in Stone

There is one type of mud that is unlike any other. It is neither grey, nor black, nor composed of minute clay minerals or exceedingly worn-down rock fragments. It is not the debris of a pulverized mountain. Rather, it is cream-coloured, or perhaps pale yellow, and made of calcium carbonate. For the most part it is thought to have derived from exceedingly eroded fragments of the hard parts of a variety of invertebrate animals, such as molluscs, crustaceans, echinoderms, corals and bryozoans, along with a smattering of calcareous algae.

Mineral dendrites, probably formed of the manganese oxide mineral pyrolusite, formed on a thin fracture in the 145–150-million-year-old Solnhofen Limestone from Bavaria, Germany.

While much of the fragmentation of these remnants of dead animals and algae is by physical abrasion, a goodly amount derives from bio-erosion. For instance, some fishes, sea urchins and molluscs graze on corals and shells, creating clouds of fine calcareous sediment. Others, including some sponges, bore into the shells of both living and dead invertebrates, creating sediment particles as they go. However, there is another source of calcareous mud whose formation does not involve the activity of animals. It may well be, though, that it is influenced by other, much smaller organisms that play an active and critical role.

One of the more intriguing phenomena that occur in oceans are what are termed 'whiting events'. These occur when the sea becomes clouded over a wide area due to the precipitation of extremely minute crystals of aragonite, less than 4 microns (μ) in diameter. Long thought to have been an inorganically induced process, there is mounting evidence that such whiting events may be due to the activity of viruses. A dominant component of phytoplankton blooms are cyanobacteria. Like all other organisms, cyanobacteria are periodically susceptible to viral infections. The effect of these intracellular viruses on the single-celled cyanobacteria can be catastrophic as the cells' walls are ruptured. This results in the release of intracellular bicarbonate ions which on contact with the seawater induce precipitation of minute aragonite crystals that in tranquil settings ultimately accumulate as carbonate mud.[23] Such extremely fine mud, lithified into hard, smooth rock surfaces, provoked one man to exploit them for a singular purpose that was to have a profound effect on printing techniques for more than two hundred years and was to give this type of limestone its own distinctive name – 'lithographic limestone'.

It all starts with a laundry list. The year is 1796 and playwright and actor Johann Alois Senefelder (1771–1834) is at the end of his tether. Born in Bohemia but living in Bavaria, he is keen to print some new

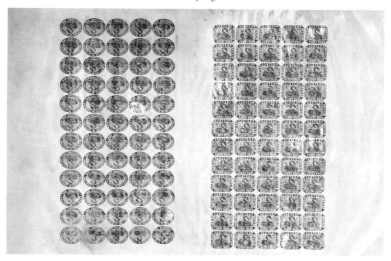

A slab of Solnhofen Limestone, used for printing Western Australian postage stamps in 1855. This particular slab was used for printing shilling and fourpenny stamps. One of the latter stamps has the famous inverted frame image of which now only fourteen remain. One recently sold for close on AU$250,000. The error is on the stamp in the right-hand block, right side, eighth stamp from top.

plays he has written. Printed they are, but Senefelder, expecting some income from their publication, finds little is left by way of remuneration after printing costs are met. There must be another way. Why not print the next batch himself? Long fascinated by the whole process of printing, he begins to experiment. First, he tries engraving on steel plates, but this is just too difficult. His engraving skills, he decides, are simply not good enough. He experiments with another material softer than steel, one he manufactures himself from clay, sand, flour and coal dust, but the cost of the equipment to print from such a template is beyond him. Then another material, softer than steel: pear wood, but the results resemble ancient woodcuts. Perhaps etching on copper would do? But he has no success here either. Or maybe he should try zinc. The result is still disappointing.

A prawn fossilized in the very fine-grained 145–150-million-year-old Solnhofen
Limestone from Bavaria, Germany. Specimen in collections of the Western
Australian Museum, Perth.

He acquires a 'handsome piece of Kellheimer stone for the purpose of rubbing down [his] colours on it'.[24] Perhaps if he wrote on
it with the wax ink he had been experimenting with it might yield
some positive results. It is then that his mother comes to the rescue.
She asks him to quickly write a laundry list down, as the laundress is
waiting. He cannot lay his hands on a piece of paper, and so he decides
to scribble the list on the piece of stone. He grabs his pen and dips it
in his homemade ink, manufactured from wax soap and lampblack
(soot), meaning to copy it down on paper later. Eventually, after paper
has been found, he starts to wipe the writing from the stone. But he
stops and begins to wonder what would happen if he etched the stone
with one part aqua fortis (nitric acid) to ten parts water. This has the
effect of leaving his writing slightly in relief after a thin surface layer of
the surrounding limestone has been etched by the acid. Any oil-based

ink applied to the wet area is simply repelled, only adhering to the original writing. Once the limestone block is inked, he can press paper onto it and obtain a perfect print. Lithography, using a fine-grained limestone soon to be called lithographic limestone, is born.

Senefelder's stone came from old quarries at Kellheim that had been quarried in Bavaria since Roman times. These particularly fine-grained limestones are now called Solnhofen Limestone, after the site where it has been quarried for hundreds of years – and continues to be – at Solnhofen in Bavaria. The environmental conditions under which the sedimentary precursor of the Solnhofen Limestone was deposited are still unclear. A number of different models have been proposed, and they all share in common the belief that the Solnhofen muds were deposited between 145 and 150 million years ago, along the northern margin of the great Tethys Ocean at a time when the climate was subtropical – warm and arid. For up to 90 metres (300 ft) of mud to have quietly and gently accumulated indicates a long period of environmental stability, probably numbering in the millions of years. The environmental setting in which this was taking place seems to have been in quiet, sheltered embayments, isolated from the open ocean by algal-sponge reefs and small, isolated coral reefs. Perhaps these were lagoons that periodically became emergent. Or maybe, they were in quiet, hypersaline back-reef lagoons.

In addition to the extremely uniform, fine-grained nature of these limestones that made them so suitable for lithography, the Solnhofen Limestone is famous for the astounding range and quality of preservation of its fossils. From birds to cuttlefish, beetles to turtles, dragonflies to fish, brittle stars to seaweed, many are preserved in exquisite detail: the feathers on the wings of the earliest bird, *Archaeopteryx*; a diaphanous cockroach wing; the whiskers on a prawn; the ink sac of a squid. The general absence of organisms that would have lived on the seafloor, combined with the undisturbed nature of the sediments, suggests a bottom environment devoid of oxygen. Animals had dropped in from the air above and from the near-surface, oxygenated waters. While the origin of the carbonate mud into which they slowly sank and which ultimately formed the Solnhofen Limestone remains an enigma, viral infections of plankton blooms, generating clouds of carbonate mud as the cyanobacterial cells ruptured, cannot be ruled out.

Turning the Tables

Despite not being on the gargantuan scale of reefs like those in the Kimberley, fossil reefs in Devon show a spectacular variation in types of limestone to such an extent that in the nineteenth century they spawned a thriving, world-class ornamental 'marble' industry. In 1851 John Woodley, from Babbacombe in Devon, was one of about ten of the leading marble masons in Britain who exhibited at the Great Exhibition in Hyde Park, London. Woodley was clearly a master of his art because, of the two prize medals awarded, one award went to Woodley for a large, round 'marble' table. It was a table constructed entirely from a bewildering array of Devonian limestones. Although not known for certain, his award-winning table may well be the one that now graces the Natural History Museum in London. Other tables also thought to have been made by Woodley are in the Fitzwilliam and Sedgwick museums in Cambridge. Each is a stunning creation made from a wide variety of limestones, all sourced either from Woodley's Petit Tor quarry at Babbacombe or from large boulders on nearby beaches.[25]

The many different limestones crafted into the tables each formed in disparate parts of the Devonian reef environment that once thrived in this part of southwest England during the Late Devonian. While the two tables held in the Natural History and Fitzwilliam museums can only be viewed from a distance because of their rarity and beauty, the other, in the Sedgwick Museum in the University of Cambridge's Department of Earth Sciences, leads a far more utilitarian existence, and has done so for at least a century. It has been a place to drink morning coffee, eat lunch and sup on afternoon tea; a place to write assignments or just to sit around and ponder the geological gossip of the day. It also allows those who sit around it to get up close and very personal with a Devonian limestone reef.

It is a spectacular table, inlaid with fifty quite different limestones, all, bar one, from in and around John Woodley's Babbacombe quarry.[26] Every part of the reef can be closely examined. No need to clamber up spiky spinifex-covered Australian limestone hills, with a coterie of bushflies in tow and the sunshine beating down. Here, in this table, almost every colour of limestone imaginable can be seen, formed in this warm, shallow Devonian sea – pink, red, green, yellow, white, grey, black, orange and brown. A wide variety of textures and fossils illustrate the various parts of the reef system in which the limestones formed. Drifting over the table is to pass from a stromatoporoid sponge muddy lagoon floor to a coral-rich patch reef, into a surge channel with storm debris, then on to the reef front upon which grew stromatoporoids and formed microbially cemented mud, finally plunging deep into the dark water off the reef front, now with equally dark grey limestones. And all the while enjoying a nice cup of tea.

Many of the 'marbles' from the Devonian quarries that were fashioned by master masons like John Woodley were red – ornate columns, altars and steps, in buildings like Exeter College Chapel, Oxford, and the Fitzwilliam Museum. So too are many of the limestones that

'Marble' table made by John Woodley in the 1850s, in the Watson Gallery of the Sedgwick Museum, Cambridge. All the multicoloured limestones are from Late Devonian reefs in the vicinity of Babbacombe, Devon, and represent limestones formed from sediments deposited in various parts of the 370-million-year-old reef system.

formed in the deep water off the front of the reef in the Kimberley region of Western Australia. In Europe these stunning red rocks have attracted people for at least the last 2,000 years. None more so than a man who decided that Devonian reef limestones were fit for a king.

Red Is the Colour

In 1669 Charles Le Brun (1619–1690), arguably the most talented artist and interior designer in seventeenth-century France, began work on designing the interior of King Louis XIV's expanded palace at Versailles. In its infancy the palace had been a small hunting lodge, but it became the king's vision to transform it into a bombastic statement of his prestige and power. Having already bestowed the title of *Premier peintre du roi* (first painter to the king) on Le Brun, Louis appointed him *directeur* of the 'Gobelins Manufactory', making Le Brun responsible for the supervision of dozens of artists and craftsmen to work on the interior decorations of the royal palaces, especially Versailles.

The most spectacular new addition to the palace was the Galerie des Glaces (Hall of Mirrors). To walk the 73-metre-long (80 yd) gallery is to be flanked on one side by seventeen huge windows, through which lie the splendours of the palace gardens. On the other side, the same view is reflected by 357 coruscating mirrors, arranged in, literally, seventeen mirror images of the windows. Between the mirrors are towering pilasters constructed from a mottled red 'marble' called Rouge de Rance. This building stone, known more generally in later usage as Rouge de Flandres, or simply Belgian red marble, was sourced from quarries in the Hainaut province of Belgium. Exploited since Roman times, where it was used in local villas, in mosaics and for other decorative purposes, this stone was utilized in large amounts for the most elegant parts of the Palace of Versailles. Not only was it one of the main 'marbles' used in the Hall of Mirrors, but it dominated the Marble Courtyard.

In other parts of Versailles, however, Le Brun preferred to use another type of 'marble'. Intricate mantelpieces were carved from a rock known as Rouge de Languedoc, from southern France. While there is more than one type of Rouge de Languedoc, the one mostly used at Versailles is called Red Griotte, taking its name from Morello cherries. This is because it contains dense accumulations of cherry red fossil ammonoids, known as goniatites, which nestle tightly together in a bright red lime mud. Their spiral shells often have white calcitic infillings, resulting in the rock being colloquially called *oeil de perdrix* (partridge eye). Both Rouge de Rance and Rouge de Languedoc, which feature so prominently at Versailles, have a number of features in common. First, like Woodley's Devon 'marbles' they are not marbles – they are limestone that formed in carbonate mud-mounds in warm, relatively shallow seas. True marble is thermally metamorphosed limestone. In Belgium, about 75 mud mounds

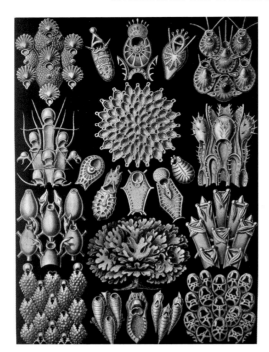

Plate showing drawings by Ernst Haeckel of modern bryozoans from his *Kunstformen der Natur* (1904).

are known, each of which is hundreds of metres in length and tens of metres high. In all likelihood these mud mounds formed due to the activity of microbial communities. Most have been quarried for use as ornamental 'marble'. Second, the Belgian and French rocks are both Late Devonian in age, that is, about 380 million years old. And third, both are red, being suffused with haematite.

Just how did the limestones gracing the palace acquire their sanguineous nature? In the 1990s Bernard Mamet and Alain Préat of the Université Libre de Bruxelles began investigating their secrets.[27] By examining thin sections of the rocks, Mamet and Préat's glimpses into this Devonian world revealed that the haematite which produces the red colour was not randomly distributed through the rocks, but was, in fact, localized. This provided one of the clues to its formation.

The colour occurs as very thin stringers of intense red that represent microbial mats formed by bacteria of 'extrapolymeric substances'. These layers of mucus are a perfect microenvironment for the activity of iron bacteria, sequestering dissolved iron from the seawater and converting it into growths of red haematite. Their activity also sometimes produces tiny millimetre-scale shrub-like microbialites called *Frutexites*. Iron bacteria only function in environments where oxygen is present, but at relatively low levels, and their activity is highest close to zones of total oxygen depletion. They also favour environments in which nutrients are impoverished. Such conditions are found within reef systems.

Cool Carbonates

Studies of the carbonate sediments that yield the many different types of limestones have been mainly carried out in warm, tropical waters, in places such as the Caribbean, the Persian Gulf, the Great Barrier Reef and Shark Bay in Australia. While it is true that these are wonderful analogues for far more ancient limestones, it cannot be denied that their locations are somewhat pleasant places in which to carry out research, as well as being readily accessible. A consequence of this has been the misguided view that limestones in the rock record all indicate warm-water conditions in the past. Nothing could be further from the truth. Some intrepid researchers have realized that many carbonate sediments are, by contrast, deposited beneath some of the coldest, roughest, least hospitable waters in the world, most notably in the Southern Ocean. Moreover, most of the best-exposed examples of limestones formed from these sediments occur in Australia and New Zealand, locations far from the centres of research into carbonate sediments in Europe and North America.

What sets these so-called cool-water carbonates apart from their warm-water equivalents is that microbial communities play little to no role in their formation. Rather, they are a shell hash, constructed almost entirely from the fragmented remains of a wide-ranging suite of marine invertebrates, all of which secrete calcium carbonate. Overwhelmingly, these cool-water carbonates are dominated by the skeletal remains of one type of invertebrate animal – bryozoans. While warm waters are the ideal location for many types of shallow-water corals, in cooler waters (less than 20°C (68°F)), the dominant colonial animals are bryozoans. Each bryozoan animal,

Bryozoal limestone of the 20-million-year-old cool-water Miocene Abrakurrie Formation, Nullarbor Plain, Western Australia, comprising fragmented bryozoans, echinoderms and bivalves.

called a zooid, occupies its own pouch-like structure, which is made of calcite. Colonies can grow large. Some, up to 50 centimetres (20 in.) in diameter, can contain up to 2 million individual zooids.

Bryozoans are extraordinary animals. Unlike corals, in which each individual in the colony is an identical clone, in a single bryozoal colony there are a variety of zooids that carry out specific tasks which support the colony in different ways. Some are specialized feeders. Others' sole aim in life is to reproduce. Others act purely as brood chambers for embryos. The feeders help these non-feeding zooids by providing them with nutrients. Others are defensive zooids, attacking and killing any larvae that try to settle on the colony. The role of some is to glue the entire colony to a substrate, such as a dead shell. There are even some, in one group called lunulitiform bryozoans, in which many of the zooids on the base of the dome-shaped colony have become adapted for mobility: they function as legs. The entire colony is able to lift itself off the substrate and 'walk'. Somehow the individuals within the colony co-ordinate their movements. Nobody has the faintest idea how they do this.

Bryozoan colonies form a wide variety of shapes that are greatly influenced by the environments in which they live. Some encrust rocks, shells or any hard substrate, often forming thin, lace-like sheets, hence their common name – sea lace. These tend to live in shallow, turbulent water, as do nodular or hemispherical forms. However, in quieter, deeper water the bryozoal colonies take on a more delicate habit, growing as thin arborescent branches or as more foliose structures. Basically, the deeper the water, and thus the lower the hydrodynamic energy, the more delicate the colony. The realization that such zoning of colony architectures can be correlated with water depth has been a boon to studies of the depth of deposition of ancient bryozoal limestones.

The most stunning outcrop of limestone in the world is a cliff that runs intermittently east–west for about 1,000 kilometres (620 mi.), forming the southern margin of the Nullarbor Plain in Australia. This is the largest limestone outcrop in the world, covering an area of almost 2 million square kilometres (770,000 sq. mi.), about 85 per cent of the area of the United Kingdom. The cliffs of limestone that face the Southern Ocean in which they were formed are climate change cliffs. The limestones of the Nullarbor Plain are composed in their lower two-thirds by bryozoal limestones. The 100-metre-high (330 ft) cliffs are striped yellow, white and red. The oldest is the 38-million-year-old Wilson Bluff Limestone, above which lies the 20-million-year-old Abrakurrie Limestone. Both are bryozoal limestones that formed in a cool sea between 10 and 20°c, when Australia was located at a higher latitude than today, having just separated from Antarctica. The uppermost, youngest part of the cliffs are formed of the Nullarbor Limestone, which contains a fossil fauna that shows it was formed in warmer waters, when Australia had drifted further north. Within the older bryozoal limestones, cycles of subtly changing proportions of the fragmentary remains of bryozoans, bivalves and sea urchins that make up the bulk of the rock tell of cycles of rising and falling seas.[28] Subtle changes in the proportions of different forms of bryozoal colonies also tell of cycles of changing sea levels. As the sea rose, so more delicate bryozoans dominate the limestones. Conversely, as it fell, the encrusting and nodular forms predominate. Bryozoal limestones are both palaeotemperature and palaeobathymetry indicators.

Clearly, over the past few billion years, limestones have been an important sink for carbon dioxide, sourced directly from absorption by oceans from the atmosphere and indirectly from bicarbonate ions derived from the chemical weathering of pre-existing limestones and

'Climate change cliffs', where the Australian Nullarbor Plain meets the Southern Ocean. Lower, white beds, middle Eocene (38 million years old) Wilson Bluff Formation, a cool-water limestone. Overlain unconformably by brown early Miocene (20 million years old) cool-water Abrakurrie Limestone. Uppermost beds, red, nodular Nullarbor limestone (15 million years old), a warm-water limestone.

cycled via river systems back into lakes and oceans. What has enabled this to happen is the activity of organisms – microbial communities, including viruses, cycanobacteria, archaea and bacteria, and a host of predominantly marine invertebrates, from corals and trilobites to bryozoans and sea urchins. The variety of limestones formed by microbes and animals is enormous, and there are still those produced by algae to come. I trust my computer is listening.

FOUR

AS CHALK IS TO CHEESE

*Chalk is a rock that we rather take for granted. Yet it is a rock of many
contradictions. White as a swan's breast. Indeed, except when it is red.
Or grey. Or sometimes green. Yellow, even. Greenish yellow as well.
Occasionally pink or violet. Blue, would you believe. And then there's the
orange variety. And rarest of all – black. So, no, not all as white as swan
feathers, nor as the cliffs that scarp the English Channel. And chalk is,
of course, soft. That's why it has been used for writing and drawing for
millennia and carved into a myriad of forms to please the eye. Except,
that is, for the chalk that is so hard it can be used to make buildings which
stand resilient for hundreds of years.*

*In its natural environment of ranging cliffs and rolling downs, chalk lies
with an altogether strange bedfellow. Black as night; so hard it splinters
into slivers when it breaks, and sharper than shattered glass. It is as cheese
is to, well, chalk: flint.*

Soaking Up the Sunshine

There are some who would have it that one particular type of rock,
coal, is little more than fossilized sunshine, seeing as how the
plants from which it formed were the direct product of photosynthesis. If that is the case – and it certainly conjures up an engaging image
– then it would only be right also to regard microbialites in exactly the
same light (so to speak). Like coal, they are the products of photosynthesis, but in cyanobacteria rather than plants harnessing the luminous
power of the Sun. When it comes to chalk, you would seemingly be

hard-pressed to give credit to the Sun for any direct involvement. Just cast your eye over the definition of chalk provided by the OED: 'A white soft earthy limestone (calcium carbonate) formed from the skeletal remains of sea creatures.' Now, sea creatures, as the dictionary further informs us, are animals; so, a rock made up of animal remains. Well, not exactly. The Sun is not to be dismissed so lightly.

* * *

CHALK IS A SINGULAR ROCK TYPE. It was of its time, only forming in a 40-million-year window, mostly in the Cretaceous period, from about 100 to 60 million years ago. It stretches across much of the world, outcropping intermittently from close to the Western Australian coast to Kazakhstan, across the Middle East and northern Europe to Britain, then over the Atlantic to North America, from Texas to Colorado and Nebraska to Arkansas. How chalk forms taxed many notable minds for centuries. John Woodward, always keen to offer an opinion on everything, is strangely silent on the matter. Being a physician, Woodward is more interested in its medicinal value, which he documents in the early 1720s in the catalogue of his collection:

> Fine clean Chalk, is one of the most noble Absorbents I know: and most powerfully corrects and subdues acrid Humours in the Stomach; tempering and allaying the Emotions and Ebullitions of them. This property is what renders it so very serviceable in the Cardialgia, or Heartburning; an Affectation of the upper part of the Stomach caused by the ascent of hot acrid corrosive Steams; and in Diarrhaeas or Fluxes . . . Those who frequent the Sea, and are not apt to vomit at their first setting forth, fall frequently into Loosnesses, which

Chalk cliffs, west of Durdle Door, Dorset.

are sometimes long, troublesome, and dangerous. In these, they find Chalk so good a Remedy, that the experienced Sea-Men will not venture on board without it. They chiefly make use of that which is contain'd in the Shells of Echini Marini; which indeed is usually very fine and pure. These are dug up very commonly in the Chalk-Pits on each side of the River, at Purfleet, Greenhyth and Northfleet, where the Chalk-Cutters drive a great Trade with the Sea-Men, who frequently give good Prices for the Shells, which they call Chalk-Eggs.[1]

In terms of how it forms, Woodward's view seems to be 'that it just is'. Being such a light (in weight) rock, chalk, Woodward believes, was one of the later sediments to settle out of the turbulent waters generated by the biblical deluge. There are other naturalists of a similar diluvial disposition, who claim that chalk is merely a natural chemical precipitate, a not-unreasonable assumption. But one man looks closer. It is the early 1830s; he peers at powdered chalk closely through a microscope and sees a bewildering variety of strange shapes. The man is German zoologist Christian Ehrenberg (1795–1876), a specialist in microscopic organisms. What his microscope reveals are countless circular and oval button-like plates of calcite, ornamented with pores or crossbars. They are tiny, just a few microns across – thousands would sit on the head of a pin. They are nothing like the microscopic animals and plants that Ehrenberg has seen before. He dismisses the idea of their being of organic origin.[2]

More than twenty years later, in 1857, Thomas Huxley, the most passionate supporter of Charles Darwin's ideas on evolution and arguably the leading comparative anatomist of the nineteenth century, examines sediment samples collected by Joseph Dayman on board HMS *Cyclops*. Dayman has been dredging sediment from the seafloor between Ireland and Newfoundland with the aim of trying to find the best location for laying a transatlantic telegraph line. Although Huxley sees a few foraminifera (single-celled protists) when he peers down his microscope, the sediment is mainly packed with what he calls a 'granular matrix'. It intrigues him. Within it are tiny, circular calcite discs, which he notes, 'at first sight, [are] somewhat like single cells of the plant *Protococcus*; as these bodies, however, are rapidly and completely dissolved by dilute acids, they cannot be organic, and I will, for convenience's sake, call them coccoliths.'[3] Huxley is looking at the same structures seen by Ehrenberg twenty years earlier in powdered chalk.

In 1861, G. C. Wallich sees coccoliths from another sediment sample from the Atlantic Ocean, but finds them assembled into minute, spherical structures, looking like baroque footballs. He calls them 'coccospheres' and concludes that they are some sort of organism.[4] The larvae of single-celled planktic foraminifera, he thinks. The same year, Henry Clifton Sorby, a geologist and microscopist, examines these little discs as he too studies a sample of powered chalk. Unlike Ehrenberg, he also thinks they are of organic origin, but not foraminifera.[5] By 1868 Huxley comes round to the view that coccoliths are indeed of organic origin, but what they are still eludes him.[6] Chalk, however, fascinates him. In the same year he gives a lecture, entitled 'On a Piece of Chalk', to 'working men' in Norwich at the meeting of the British Association for the Advancement of Science, and published it in *Macmillan's Magazine*. He observes that

the chalk is no unimportant element in the masonry of the earth's crust, and it impresses a peculiar stamp, varying with the conditions to which it is exposed, on the scenery of the districts in which it occurs. The undulating downs and rounded coombs, covered with sweet-grassed turf, of our inland chalk country, have a peacefully domestic and mutton-suggesting prettiness, but can hardly be called either grand or beautiful. But on our southern coasts, the wall-sided cliffs, many hundred feet high, with vast needles and pinnacles standing out in the sea, sharp and solitary enough to serve as perches for the wary cormorant, confer a wonderful beauty and grandeur upon the chalk headlands.

However, it is not until 1872 that the affinities of these tiny discs seen in powdered chalk and deep-sea sediments are finally settled,

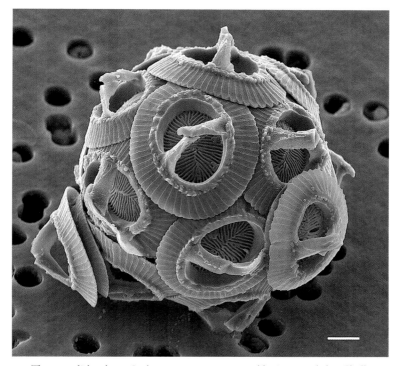

The coccolithophore *Gephyrocapsa oceanica* encased by its coccoliths. Chalk sediment is largely composed of such coccoliths.

when samples floating in the ocean are collected by John Murray on the pioneering oceanographic expedition of HMS *Challenger*.[7] He realizes that the coccospheres made up of an encircling covering of plates are a particular type of single-celled calcareous algae. Today they are classified as photosynthetic haplophyte algae, and known as coccolithophores. Stimulated by light, tiny vesicles form within the coccolithophore and produce the coccoliths. These progressively move to the outside of the cell, replacing older coccoliths, some of which are shed into the water. And so, a multi-layered spherical shell is formed: the coccosphere. The coccoliths seen by Huxley, Wallich and Murray in their oceanic sediment samples are derived from the self-same type of organism of which chalk is composed.

Like coal, and microbialites, chalk is the product of sunshine. However, turning an incalculable number of single-celled algae into chalk rock was obviously a complicated process. A concatenation of factors had to come into alignment in time and space.

Into the Depths

For tens of millions of years during the Cretaceous Period, sneaking into the beginning of the following Palaeogene Period, one-third of today's land surface was covered by the sea. In its depths it slowly, but resolutely, accumulated fine, white chalky mud. Sea levels were much higher during those times, not only because Earth's climate was so warm that few, if any, ice caps existed at the poles, but because seafloors were being actively constructed. What this meant was that extrusion of molten rock over the seafloor displaced the sea upwards, pushing it onto what had once been dry land. This was a period of active spreading of the continents. The Atlantic Ocean was still being created as the American continents drifted away from Africa and Europe. Moreover, the great supercontinent of Gondwana was in the final throes of ripping itself asunder. Africa, India, Madagascar, Antarctica and Australia were all splitting apart from each other to become discrete lands.

The consequence was twofold: the upwelling of vast amounts of magma, creating a new seafloor and raising the sea level, but also the production of vast amounts of gases that belched into the atmosphere, in particular carbon dioxide. (It has been estimated that atmospheric carbon dioxide levels during the Cretaceous were up to six times higher than they are today.) It was, therefore, a hot

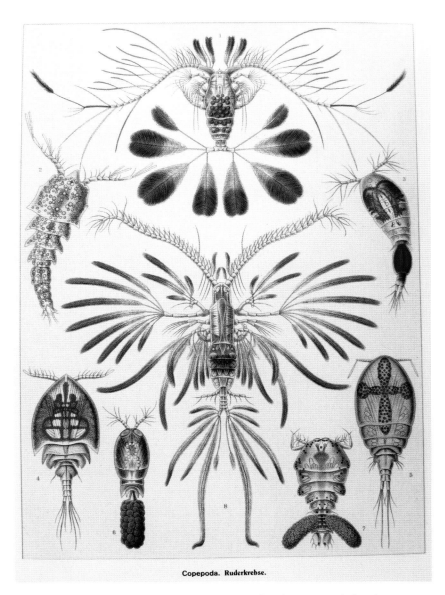

Copepoda. Ruderkrebse.

Plate showing drawings by Ernst Haeckel of modern copepods from his
Kunstformen der Natur (1904).

world. For coccolithophores, these conditions were manna from heaven – more shallow seas into which to replicate and, for these photosynthetic phytoplankton, greater quantities of carbon dioxide dissolved in the oceans to absorb and turn into calcium carbonate to feed the sediment. With the higher sea level there was much less land to erode, so far less silt, sand and mud to be transported into the oceans. This was the ideal setting for the formation of limestone – in this case, chalk. For about 40 million years, hundreds upon hundreds of metres of fine chalk mud, composed largely of coccoliths, silently accumulated.

The Cretaceous and early Palaeogene seas in which chalk was being deposited were probably no more than about 300 metres (980 ft) deep.[8] Being photosynthetic, the coccolithophores were confined to relatively shallow water – a zone between about 20 and 80 metres (65–260 ft). Any deeper and there was insufficient light for photosynthesis; any shallower and there was just too much. Despite being such a dominant part of the oceans' phytoplankton, coccoliths are something of an enigma. Nobody really knows what their function is within the coccolithophore, but many suggestions have been made. Perhaps they help shield the cell from sunlight that is too intense in shallow water; or maybe they are just a metabolic by-product of photosynthesis – the result of detoxification of unwanted calcium carbonate. They could also be support structures for the cell, or even defensive armour. Another possibility is that they function as light-gathering or light-concentrating devices. Alternatively, their weight could help provide the cell with improved stability. One thing is for certain about how they form: we simply don't know. The reality may well be that they serve multiple functions.

When coccolithophores die the individual plates spall off the coccosphere, but being so small and so light, they barely sink. The coccosphere is essentially broken into its component coccolith parts in two ways: either by exploding apart following infection by a virus, or by being eaten by zooplankton, usually tiny crustaceans called copepods. When coccolithophores rupture following viral infection and the isolated coccoliths are released, their subsequent fate can be either to sink, to dissolve or even to fly – to become elevated into sea-surface aerosols and be carried into the atmosphere. It has even been suggested that following viral infections of massive algal blooms, some of which can reach up to 100,000 square kilometres (38,000 sq. mi.) in area (about the size of Egypt), enough coccoliths can be released into the atmosphere that they become seeding agents of water droplets in clouds. For the vast majority of coccoliths that do not bring rain, they are likely to be dissolved in oceanic water rather than slowly drifting down to the seafloor. Even if somehow a single coccolith avoided such a fate, it has been estimated that it would take it about ten years to sink 300 metres (980 ft). There is, however, a much faster way for coccoliths to descend into the dark oceanic depths, and that is by passing through the gut of a rather hungry copepod.

One of the most common biominerals in the ocean is the calcite produced by coccolithophores. The nature of the chemistry of the oceans is such that calcite should not readily dissolve in waters above a depth of 4,500 metres (14,760 ft). Yet it has been estimated that up to 80 per cent of calcium carbonate is lost in the upper 500–1,000 metres (1,640–3,300 ft) of oceans.[9] The site of much of this dissolution is very specific; it is in the guts of zooplankton, in particular copepods. These crustaceans are tiny. If you could keep it still, a single copepod would sit comfortably on the head of a pin. Experiments have shown that if the little crustaceans are starved for a while, then allowed to binge on coccolithophores, up to 38 per cent of the ingested calcareous coccoliths would be dissolved in their guts. However, as they continue to feed, the copepods' guts become buffered, and there is no more dissolution.[10] The coccoliths just start to accumulate. The authors of a study call this, somewhat whimsically, the 'Tums hypothesis', named after the over-the-counter remedy for excess stomach acid. With less dissolution in their guts, the copepods need to get rid of the coccoliths, so they void them as faecal pellets. A single pellet may contain tens of thousands of coccoliths. Their weight is fairly substantial, so pellets will fall through a 300-metre (980 ft) column of ocean water in a couple of days, accumulating as the carbonate ooze that coats the seafloor – the precursor to chalk rock.

Thus, for about 40 million years, mainly during the latter part of the Cretaceous Period, it rained copepod poo. In fact, it was such a prolonged deluge that it saw the accumulation of hundreds of metres of copepod excrement. Accumulating on the seafloor, this chalk slurry was about 98 per cent pure calcium carbonate. While most of the individual carbonate fragments in chalk are coccoliths, there are also a few other components: single-celled protistan foraminiferans (a substantial part of the oceanic plankton), minute prismatic fragments of the bivalve *Inoceramus* and the fragmented debris of sea urchins and other echinoderms.

During the Cretaceous Period, the chalk seafloor was topographically very uneven, with countless basins and troughs. Consequently, it was affected by bottom currents that frequently reworked and redistributed the chalk ooze. Moreover, the rain of copepod faecal pellets and other calcareous plankton to the seafloor was not constant. When the plankton rain became a drizzle, or ceased altogether, there was reduced sediment accumulation and the seafloor became a thriving environment for burrowing invertebrates, such as lobsters and sea urchins. They left evidence of their activity in extensive, often labyrinthine, burrow systems. If reduction in sediment accumulation was prolonged, then the upper layers of chalk ooze began to undergo some changes. Down to a depth of about 1 metre (3 ft), the sediment could begin to undergo cementation – sediment turning into a rock. Chalk had begun its long journey from soft ooze to the rock that today fronts the cliffs of the English Channel. Starting at scattered points within this upper metre of chalk ooze and spreading outwards, the cementation slowly hardened and compacted the ooze into localized nodules of chalk. The area between the harder nodules remained soft, so lobsters, sea urchins and other burrowers could continue to construct their elaborate burrow systems. But as the nodules continued to grow even larger, the burrows in the soft sediment became increasingly constricted and distorted. Periodic currents could winnow this away, leaving a hard, pebbly chalk pavement. If the rate of planktonic rain remained low for a prolonged period, from tens to even hundreds of thousands of years, then eventually the nodules merged to form what is appropriately called a hardground.

Another world developed on this dark, chalky seafloor; a whole new ecosystem became established, dominated by specialist borers and encrusters. Sponges and bivalves bored into the new, hard chalk rock, while a variety of encrusting organisms took up residence on the surface. Serpulid worms encrusted the hardground with their calcareous tubes. Colonial bryozoans coated the new, hard surfaces with their lace-like structure, accompanied by an assortment of oysters. Over prolonged periods the hardgrounds sometimes became even more strongly cemented by phosphatic minerals. With a renewal in planktonic rain, the hardgrounds were eventually buried beneath new, softer chalky sediment, and the process could start all over again. The rhythmic cycles of hard and soft rock continued on their chalky way.

Rhythm of the Rocks

Viewed from the deck of a cross-channel ferry, chuffing its way from France across the Straits of Dover, it would be easy to imagine that the towering white cliffs are gaining height with each passing wave. On arriving in Dover, should any of the passengers take themselves off to a beach at the base of the cliffs and peer closely at the rocks, they might be surprised to find anything but a monotonous wall of pure, white chalk. Instead, they would find a cliff showing a variety of rhythmic cycles in the rock. Not only might they see the hard and soft layers, alternating over and over again, reflecting differential degrees of cementation of the chalk ooze, but there would be thin beds of white and not-so-white chalk taking turns. Even higher in the cliff they might see repeated thin bands of a rock as different from chalk as cheese.

The widespread deposition of chalk sediments started about 100 million years ago, at a time when pterosaurs winged their way through the skies, reptiles the size of killer whales cruised the oceans and a menagerie of dinosaurs of all sizes owned the land. In the cliff sections in both the south of England and Yorkshire, where the cliffs abut the North Sea, the lower part of the chalk sequence consists of alternating layers of white and grey chalk. This grey chalk (known as marl) gained its colour from the greater proportion of clay than within the white layers, comprising up to 30 per cent of the sediment. These cycles in the purity of the chalk are not local phenomena. Studies of these beds in the Isle of Wight and about 3,000 kilometres (1,800 mi.) to the east in Crimea reveal exactly the same sequence of

Lower chalk, Isle of Wight, showing Milanković-driven banding of chalks with variable clay content.

bundles of cycles.[11] This not only allows a close correlation between individual beds in the two regions to be made, but it supports the contention that these cycles are the outcome of astronomical forcing – perturbations in Earth's orbit.

These individual rhythmic couplets of white and grey chalk, generally 20 to 50 centimetres (8–20 in.) in thickness, are believed to represent the so-called 25,000-year precession cycles. What the variation in percentage of clay represents is less clear. Presumably it is a climatic signal, possibly periods of greater rainfall, resulting in more fine mud being washed from the land into the chalk sea than during drier times. Each of the cycles is bundled generally into groups that are thought to reflect 100,000-year eccentricity cycles.

Such cycles may have controlled not only the colour of the chalk but the lives of humans.

* * *

PEOPLE HAVE LIVED IN BRITAIN for a very long time. Nearly 1 million years, to be imprecise. But their presence in our current Ice Age world has been erratic, as the cycles that controlled the waxing and waning of glaciers across the country have also controlled the fate of human lives. For the last 600,000 years or so, these warming and cooling cycles have occurred with a roughly 100,000-year periodicity. Prior to that the glacial/interglacial cycles seem to have been influenced more by shorter-term, roughly 40,000-year, cycles. The effect of this profound cyclical climatic variability for 1 million years, from warm temperate to cold glacial conditions, was that people in Britain came and went with almost metronomic precision – about nine times over the last million years, as the ice exerted then released its frozen grip. During the warmer interglacial periods, when the glaciers receded to the north and the country began to become more habitable, people spread north from southern Europe, accompanied by a diverse range of fauna and flora. Some of the more exotic of their accompanying menagerie included hippopotamuses, rhinoceroses, lions and elephants. The evidence for the presence of humans in Ice Age Britain comes, in part, from rare skeletal remains, but mainly from their discarded stone tools and detritus formed during the tools' manufacture.

During one of the short, warmer episodes that punctuated the last cold phase, about 40,000 years ago, there was immigration into Britain when *Homo sapiens* made its first foray across what is now the English Channel, but which then was dry land. Before then, other archaic humans had been periodic visitors at times of low sea level. In the interglacial periods from about 200,000 years to about 500,000 years ago, one of the more distinctive stone tools that mark human presence in the country are the same types of hand axes found in million-year-old sites in banded ironstone country in southern Africa. These tools mark the presence in Ice Age Britain of archaic human species, probably *Homo heidelbergensis. Homo neanderthalensis* also left evidence of their periodic occupation in skeletal remains and characteristic stone tools.

Known as Acheulian hand axes, named after characteristic hand axes found at Saint-Acheul in France in the late nineteenth century, the manufacture of these tools is generally considered to be one of the most important advances in the evolution of human cognitive development. Their creation demonstrates the evolution in early humans of a fundamental awareness and appreciation of symmetry. More than that, they reveal the development of an ability to fashion irregular-shaped objects extracted from the environment into not only useful bilaterally symmetrical objects, but items that were aesthetically pleasing.[12] Tuck your thumb into your hand, fingers close together, and this approximates an Acheulian hand axe in shape and size. Both sides are knapped to razor-sharp edges. Being able to kill and butcher with something that you delighted in making, and which basically just looks good, must have been a satisfying feeling.

Bifacial Acheulian tools found at archaeological sites at Stoke Newington in London, principally in the late nineteenth century by Worthington G. Smith, occur in gravels formed in the interglacial period.[13] These are correlated with what is known as marine isotope (MIS) stage 9, which is dated at 300,000–337,000 years. The famous site at Swanscombe in Kent, which has yielded thousands of hand axes, correlates with the even earlier MIS stage 11, which occurred about 400,000 years ago. Similar stone artefacts found at an archaeological site at Boxgrove in West Sussex occur in sediments deposited in yet older MIS stage 13, about 500,000 years ago. Even more ancient deposits, at Pakefield in Suffolk, are 680,000–750,000 years old, and the oldest of all, at Happisburgh in Norfolk, dating to 850,000–950,000 years, has yielded stone artefacts, perhaps made by *Homo antecessor*. This evidence all attests to humans periodically moving north into Britain with these climatic changes.

The stone tools made by various human species were principally crafted from a type of rock that was easy to find in much of southern and eastern Britain, and ideal for such a use: chert. However, there is chert, and there is chert. The form frequently used by tool-makers in prehistoric Britain was very distinctive, being usually coal black, though it could sometimes be pale grey or brown. Its origins lay in the chalk deposits that coated much of southeastern England. For at least 1,300 years, this type of chert has been known as flint. And, in all likelihood, it also owes its origins, some 70 million years earlier, to much the same Milanković cycles that also controlled the human occupation of Britain.

As well as being able to pluck flint from chalk outcrops, our archaic predecessors would have been able to source it from a variety of other places. One was the beaches lining much of southeast England, where water-worn pebbles of flint crowd the shoreline; another was along the banks of rivers. Flints have been eroded out of the chalk by rivers that have periodically coursed across the country, particularly during warm, wet periods following the retreat of the glaciers. The rivers left behind gravels dominated by rolled cobbles of flint – easy pickings for someone looking to turn a rock into a tool. Like other forms of chert, flint was ideal. Hit it with another stone and slivers with edges as sharp as the finest scalpel would fly off. These flint splinters could be hafted to a stick, making an effective killing weapon. The core that was left behind was ideal for fashioning into a hand axe to serve a multitude of uses, from scraping, digging and chopping to killing. What these ancient tool-makers are most unlikely to have realized was that the rocks upon which they relied for their very existence originally existed in quite another form – as vase-shaped animals anchored to an ancient chalky seafloor, quietly filtering the ocean water. These animals were sponges.

You could, perhaps, be forgiven for assuming that the genesis of a rock that is so well known, having been used so much and for

Claude Monet, *The Cliffs at Étretat*, 1885, oil on canvas.

so long – as tools for thousands of years, and as building stone for dwellings for our bodies and souls – had been well established. Such an assumption, I fear, would be misguided. How black flint came to be interspersed with bright, white chalk has perplexed many minds for a long time. The secret to its formation has been largely unravelled in recent years and found to lie in a number of key features. The first of these is that rather than being haphazardly scattered through the chalk, flint occurs generally in a rigidly defined manner – as layers, set between about half a metre and 2 metres (20–80 in.) apart. The flint can occur as flat, tabular, continuous beds that extend over many kilometres, or as irregular nodules in discrete layers. Each flint layer is usually quite thin, often little more than a hand's breadth in thickness. There are, however, some exceptions to this layered rule, and these provide another key element in helping to unlock the secret of flint's formation. Most notable are the flint nodules that are vertically, rather than horizontally, orientated in the chalk. These are interpreted as the infills of what were probably once burrows made by a variety of invertebrate inhabitants of the chalk sediments, such as crustaceans, bivalves, sea urchins and worms. The vertical components of their burrows appear to have been preferential conduits within which flint formed. What this shows is that flint was not formed contemporaneously with the chalk sediment, but some time after.

The towering extent of cliffs of alternating layers of white chalk and black flint is often impressive. While the iconic cliffs of Dover loom large in the English psyche, there is another impressive set of chalk cliffs situated on the other side of the English Channel in Normandy, at a place called Étretat. Here the cliffs have been made famous not by mythical bluebirds flying over them, but by virtue of being the source of artistic inspiration for a whole gamut of late nineteenth-century French artists, among them Claude Monet, Eugène Delacroix, Henri Matisse, Édouard Manet, Eugène Boudin, Jean-Baptiste-Camille Corot and Gustave Courbet. Monet was particularly fascinated by the Étretat cliffs at one locality – Porte d'Aval – being inspired to paint them more than fifty times. As well as the cliffs' structure, with their natural arch and stack, and their apparent change in colour as the Sun moved across the sky, Monet's imagination seems to have been piqued by the vivid stripes of flint that figure prominently in his paintings. Rising about 90 metres (295 ft) out of the sea, the chalk beds here are punctuated by more than one hundred layers of flint. Like the cycles of pure chalk and chalk marl, and the cycles of soft and hard chalk in the lower chalk beds, these alternating bands of chalk and flint, which extend laterally over great distances, are also the product of astronomically driven forces.

Cyclical chalk and flint bands that characterize many of the chalk deposits in Europe have been studied in detail in the Netherlands – in the upper part of the latest Cretaceous Gulpen Formation, which is some 68 million to 72 million years old.[14] Deposition of each chalk and flint cycle would have taken thousands of years. The existence of roughly 75 of these alternating beds strongly suggests the influence of Milanković cycles, with each flint bed equating to 20,000-year precession cycles. These flint layers are bundled roughly into groups of five, equating to eccentricity-driven cycles of approximately 100,000 years. These are further grouped into individual sequences of around twenty layers, equating to roughly 400,000-year eccentricity cycles. Again, we see Earth's orbit around the Sun and the irregularities of its axial spin influencing the amount of solar energy being received, and therefore the nature of our planet's climate. Just how variations in the amount of solar energy reaching Earth influence the timing of flint formation, however, is still not fully understood. Exactly what it

was about the chalk seas that triggered its formation at such regular intervals is also something of an enigma.

Stretch your mind and imagine being able to cruise above a chalk seafloor, some 300 metres (985 ft) below the surface of the sea. Light would struggle, then fail, to penetrate to this depth. It would therefore be a pitch-dark world. Ironic, somehow, for such a white, bright sediment to be forever in the dark. Suppose, however, that you are able to shed some light on this murky scene. And, furthermore, should your submarine visit fall during one of the flint-forming periods that only comes around every 20,000 years or so, would you see a seafloor littered with globules of black, silica gel in the process of crystallizing into flint? Or could it be that, irrespective of how many times you visit, you will only ever see a seabed covered in white mud, with not a flint nodule in sight? For, if so, then the secret to where flint forms may not be found where sediment and ocean water meet, but rather hidden deep within this accumulating pile of white copepod dung.

The key to unlocking this secret is found, not surprisingly perhaps, in the structure and configuration of the nodules themselves, particularly their relationship to the chalk within which they are entombed. It has long been observed that many flint nodules appear to be infills of burrows made by a range of different invertebrates that often meandered in a labyrinthine fashion through the chalk. Most spectacular are those that were first described from chalk deposits in Northern Ireland in 1817 by the Reverend William Buckland. These are vertical funnels of flint, like enormous sausage rolls, some up to 9 metres (30 ft) in length, with a chalk filling and flint pastry, ploughing vertically through the sediment. These structures Buckland called

Upper chalk, Isle of Wight, showing tilted bands of chalk and black flints. Note Professor Andy Gale for scale.

'paramoudra'.[15] They have since been found in a few other areas, such as chalk deposits in Norfolk. Paramoudras are now generally accepted to have once been very large vertical burrows. Their verticality clearly demonstrates that this type of flint did not form at the same time as the chalk was being deposited. It came later, as the chalk was being turned into rock – a process called diagenesis.

As the hardgrounds in the chalk show, it didn't rain planktonic debris in the chalk seas continuously. There were regular periods, corresponding to Milanković precession cycles, when it stopped raining and sediment failed to accumulate. But sometimes, rather than the chalk being cemented into hardgrounds, flint formed. Why this happened is not clear. Whether it was climate change affecting temperature, or maybe nutrient supply being disrupted, adversely affecting the growth of the calcareous plankton that fed the chalk sediment, is not known. During these quiescent periods anaerobic bacteria were busy, processing and destroying organic matter in the upper metre or so of sediment. As a result, oxygen levels in the sediment were progressively reduced, especially deeper in the sediment. Here was to enter an oxygen-deficient world where another completely different suite of anaerobic bacteria flourished. The boundary between these aerobic and anaerobic worlds was very narrow and remained at a fixed level during the cessation of sediment accumulation. Communication between these two bacterial environments also took place via various conduits, especially the burrows created by crustaceans, sea urchins and molluscs.

In the critical zone between the two bacterial worlds, oxygen in the upper reaches of the sediment layer combined with hydrogen sulphide produced by anaerobic sulphur-oxidizing bacteria, to generate hydrogen ions. The effect of this was to simultaneously dissolve the sedimentary calcium carbonate and liberate hydroxyl complexes.

Flint nodule from the Isle of Wight formed around a crustacean burrow. Organic matter within the burrow probably facilitated the replacement of chalk by silica.

These acted as catalysts for the precipitation of either silica gel or a hydrous form of silica called opal-CT.[16] The source of this dissolved silica was mainly dead sponges that had periodically forested the seafloor. Many types of sponges have an internal skeleton made up of a meshwork of siliceous spicules that eventually dissolve after the organism's death. This silica source was augmented by other dissolved silica derived from a range of planktonic organisms, especially siliceous algae (diatoms) and single-celled protists (radiolarians). The hydrous silica precipitated by the interaction between the two communities of bacteria sometimes accumulated as nodules within the labyrinthine burrow systems. At other times it simply replaced the chalk in horizontal layers at the boundary between the oxygenated and non-oxygenated zones, forming just a few metres down in the sediment. Once sediment production resumed with the blooming of coccolithophores, this static boundary was disrupted, and formation of silica gel or opal-CT ceased. Over time, these nodules or layers of silica that had developed in the sediment crystallized into microcrystalline quartz. And so, flint was formed. Sponges, suites of bacteria and wobbles of Earth's axis had therefore all come together to produce the raw material that enabled archaic species of humans to stretch their imaginations and, for hundreds of thousands of years, craft a mighty cache of life-enabling tools.

Seeing Red

Through his artworks, Claude Monet was able to transform the chalk at Porte d'Aval into a kaleidoscope of colours, aided by the ever-changing light throughout the day, creating striped cliffs of yellow, green, grey, violet, blue, pink, red and sometimes even white. But the chalk seas had done this long before Monet, conjuring up their own dazzling palette of coloured chalks. The one that somehow seems to be the most incongruous is red chalk.

The little town of Hunstanton in Norfolk is a rather strange place. It is located on the east coast of England, like all Norfolk coastal towns. And, like them, the North Sea laps at the foot of its cliffs when the tide is high. But despite this, as the Sun sets in its usual place in the west it somehow manages to shine on the chalk cliffs that rise from Hunstanton's beach. Stranger still is the fact that these chalk cliffs are not simply as white as the driven snow; they are, in part – especially when illuminated by the setting Sun – the colour of fresh blood. The upper 2 metres (6½ ft) of the cliff are the usual white chalk, but the lower 2 metres to the beach are a red-brown sandstone overlain by red chalk, making the entire edifice resemble an enormous Polish flag. The white and red chalks do not gently merge one into the other, but rather pass with an almost brutal abruptness that signifies that a major environmental change occurred between the two chalks.

Such bands of coloured chalks occur widely on the land surface in eastern England and northern Germany, and subsurface beneath the North Sea. The thickest layers, like those at Hunstanton, are about 100 million to 105 million years old, although thinner horizons are known in younger rocks, up to about 70 million years in age. These coloured chalks mainly range in hue from dark red through to pale purple and pale pink, and pale yellow to reddish brown.[17] Their variable colours are all produced by two oxides of iron: haematite, in the case of the red, pink and purple chalks, and the hydrated iron

Red chalk, Hunstanton, Norfolk, overlain by white chalk.

oxide mineral goethite, in the yellow and brown chalks. There are even bluish grey chalks in some cliff sections associated with the red chalks in Speeton, North Yorkshire, deriving their colour from abundant amounts of the iron sulphide mineral pyrite. Most common of the coloured chalks, though, and the most spectacular, must surely be the red and pink chalks. Especially intriguing is just how this chalk, unlike the white chalks, ever came to be infused with haematite. The search for their secret history lies not in the blood-red cliffs of Norfolk and Yorkshire, however, but in the wooded country of the Ardennes in Belgium, in the thyme- and marjoram-covered hills in Languedoc, southern France, and back in the Palace of Versailles.

While the Palace of Versailles may seem, in all regards, very distant from the north Norfolk coast, the same bacterial process responsible for the production of the red 'marbles' that Charles Le Brun was so keen to use also made the red chalks. At the abrupt junction between the red and overlying white chalk at Hunstanton is a crusty layer of the self-same microbialite – *Frutexites*. Like the red Devonian rocks in France and Belgium, and their counterparts in the Western Australian Devonian Great Barrier Reef, the red chalks in all likelihood owe their existence to the activity of iron bacteria, thriving on and within a seabed deprived of much oxygen. But with an abrupt change to more oxygenated water, 'normality' was resumed as the chalk returned to its usual unsullied whiteness.

It's Green, They Say

William Smith's use of colour in his geological maps was a masterstroke: yellow, for yellow Jurassic limestones; red for 'Red Marl'; black for 'Coal Measures'. But when it came to chalk, Smith chose to colour it green, not white. He could have left it uncoloured, but no.

Green it was. Was Smith influenced, perhaps, by the fact that in the early nineteenth century, chalk downland was often covered by bright green grass? He may, however, have had another reason – the fact that chalk can sometimes be green.

Throughout most of the lands where Cretaceous chalk forms the underworld, it overlies a rock type known as greensand. This consists of quartz sand grains interspersed with a green mineral called glauconite. An iron potassium silicate with a platey structure, it is a type of mica characterized by its striking green colour. However, glauconite is not confined to greensand. The transformation from the sedimentation of quartz sands and greensand to chalk was a dramatic event in Earth history. It was as though the calcium carbonate switch had been turned on and coccoliths began to be formed in monumental numbers, raining down onto the seafloor. Early in this event, the production of glauconite failed to stop, even though the provision of quartz sand did. Earth had become drier and warmer as rivers dwindled and failed to introduce the quartz sand, and carbon dioxide levels were rising rapidly at this time, pushing up global temperatures. The coccolith-producing algae began soaking up the CO_2 and, with the aid of copepods, converting it to chalk sediment, and in the faecal pellets that came out of the coccolith-eating copepods and accumulated on the seafloor lay the source of the green glauconite.

The carbonate chalk sediment that accumulated on the seafloor did so in a generally low-oxygen environment. The consequence was that bacterial decay of the organic matter in the faecal pellets generated minute crystals of green glauconite. The tiny calcium carbonate shells of the single-celled protists, foraminifera, also comprise an appreciable component of chalk sediments. Glauconite sometimes formed within their shells with the bacterial decay of their soft tissue. As chalk sediment became thicker, the glauconite content dwindled because the environment became more oxygenated, and the chalk changed from pale green to white.

These green chalks were not confined to the early formed chalks of southern England and northern Europe. Far, far away, at the easternmost extent of the Tethys Ocean, the great chalk sea that extended from northern America halfway across the world to Australia, is the Gingin Chalk. Outcropping in the odd quarry or eroded gully in southwest Western Australia, it is barely 30 metres (98 ft) thick. Yet, throughout its entire modest thickness, it is a pale-green chalk, due to the continued presence of glauconite, and, like southern England, it is underlain by greensand, the slightly disturbingly named Poison Hill Greensand. Not only is there this striking similarity in the sequence of rocks separated by a distance of nearly 15,000 kilometres (9,300 mi.), but the same species of fossils can be found in Western Australia as can be found in Texas, Sussex and the Crimea. These seemingly inconsequential fossils are important in that they tell Cretaceous time.

Aside from his geological maps, William Smith's other great claim to fame was his realization that fossils could be used to correlate strata: if the same species of fossil occurred in rocks in two geographically separated locations, then it could be argued that the sediments in which they were preserved were both deposited at the same time. And so the science of biostratigraphy was born. The shortest biostratigraphic unit of time is the 'zone' or 'biozone'. Zones are clustered into 'stages', and these into geological periods, like the Cretaceous.

Classic examples of zone fossils occur in the chalk. They are species of the echinoderm group, the crinoids or sea lilies. One, a golf ball-sized crinoid covered by large polygonal plates, lived for a short period of time, but was virtually cosmopolitan in its distribution – ideal attributes for a zone fossil. Its name is *Marsupites testudinarium*, and it appeared in the fossil record after another globular crinoid covered in small plates called *Uintacrinus socialis*. The *Marsupites* zone marks the top of the Santonian Stage. Another species, *Uintacrinus anglicus*, then replaces *Marsupites* and marks the beginning of the overlying Campanian Stage. What is remarkable is that these three zone fossils occur in the same juxtaposition from dusty chalk outcrops in Texas, to towering cliffs on the Isle of Wight, to exposures in the Crimea and eroded gullies in Western Australia. There is, however, one difference between the Gingin and other distant chalks at this time: it remained green. The others were as white as, well, chalk.

* * *

WITHOUT LIFE THERE would have been no chalk. Without carbon dioxide there would have been no chalk. Without the Sun – well, you get the idea. And without chalk there would have been no flint. For the cycles of hominids who roamed Europe, moving with the ever advancing and receding glaciers over the last half a million years or more, flint was essential – the number-one tool that kept them alive. And it all started with single-celled algae being consumed by tiny crustaceans and their voided carbonate remains becoming enormous faecal piles on the seafloor. Upon this, sponges periodically lived, died and dissolved. Then helped by anaerobic bacteria, at certain horizons and certain times they replaced the pure white chalk and transformed it into hard, black flint. Controlling the phases of chalk and flint were the forces of Earth itself – the itinerant wobbling on its axis and its eccentric behaviour as it languidly circled the Sun. The rhythmically pulsating fluxes in solar energy they created dictated changes in climate and in the pattern of life. The life on Earth that made rocks.

FIVE

A BREATH OF FRESH AIR

The colours are stunning. Not just the rocks but also the vegetation – and the sky. Three colours dominate. At the bottom, ridges and low cliffs of intense red, iron-rich rocks that structure this ancient landscape. Above, the rich green of the spiky spinifex grass. And higher still, for most days of the years, a cobalt blue sky. Each seemingly so unutterably different from the other. But in the story to be told, all three are intimately interwoven with a common heritage – of sunlight, ancient oceans and the air we can now breathe.

Air Apparent

Limestone might be thought of as the ultimate rock of ages, but there is another, very distinctive rock type that only formed in Precambrian times, and even then, for the most part, primarily in a relatively narrow window of time, between about 2.8 billion and 2.4 billion years ago. Apart from small deposits that formed under strange circumstances about 650 million years ago, you can search high and low from mountain peak to valley floor across the world in younger rocks and you will never find any. Yet, in rocks of the right age, they form enormous deposits sometimes more than a kilometre thick on every continent on Earth (apart from Antarctica), in Australia, India, the United States, Canada, Russia, South Africa, Brazil and Ukraine. Not only are they ostentatiously flamboyant rocks formed of alternating layers of red, grey and black minerals, but economically they are one of the most important natural resources found on the planet.[1] Without them our supplies of cars, washing machines and even baked-bean cans would be severely compromised. From chemically generated iron-rich sediments, these rocks, called banded ironstones, were formed. And they owe their existence, in all likelihood, to the evolution and diversification of life.

The world into which banded ironstones came into existence was quite unlike the world now. For one thing, the composition of the atmosphere was entirely different from what we experience today. The effect of this was profound: the way rocks weathered; the colour of the sky; even, it has been suggested, how fast raindrops fell. All, it has been claimed, different from today. So too the seas, whose chemical nature was quite unlike that of modern oceans. It was another world, but one, nevertheless, whose mysteries can be unravelled by unlocking secrets hidden in some of the oldest rocks on the planet. This early period of Earth history, the Archaean Eon, during which banded ironstones first appeared, extended from 4 billion to 2.5 billion years ago and takes its name from the Greek words *archaeos* and *aion* that basically mean 'ancient age' – quite appropriate for a period so far back in Earth's history that it is really pointless even trying to imagine how long ago it was.

From Earth's birth, about 4.56 billion years ago, to the Archaean Eon, half a billion years later, it is likely to have been literally 'hell on Earth', which is perhaps why it is somewhat aptly called the Hadean Eon. Although the time of transition from the Hadean to the Archaean at 4 billion years might seem an arbitrary figure plucked at random from the air, it actually coincides (more or less) with the end of the so-called Late Heavy Bombardment. Peaking between 4 billion and 3.8 billion years ago, this represents a rather hostile period in the history of the Solar System when Earth (and the Moon) were subjected to a cosmic assault the likes of which our planet had not experienced since the Moon itself was ripped from Earth by a spectacular collision, half a billion years earlier. This cataclysmic event had particularly long-term ramifications, the resultant tilting of Earth's axis to 23° or so giving the planet its seasons. This seasonality has greatly influenced the formation of a number of rock types.

During the 'Late Heavy Bombardment', large numbers of gigantic asteroids and comets collided with the young Earth. Peer at the Moon through a telescope and you will see the products of this bombardment – massive craters. Why do you see them on the Moon and not Earth? Simply put, because Earth is a dynamic, evolving planet; its surface is battered by wind, rain and the Sun, remorselessly grinding the rocks away and, on a larger scale over huge tracts of time, sucking them down into the molten Earth to be recycled. The Moon is dead. Earth lives.

Given that, since the early 1970s, it has been generally accepted that for much of Earth's history the continents have floated around in a generally placid and stately manner, any attempt to reconstruct the positions of continents and oceans in the deep time of the Archaean Eon is fraught with many challenges. The evidence is nebulous: very few rocks from the first billion years of Earth's history still exist, and those that do have mostly been extremely altered.

The heart of the Pilbara region in Western Australia is one of the most ancient cratons. These are very old and stable pieces of Earth's two outer layers – the crust and the upper mantle. Coming into existence more than 3.5 billion years ago, the rocks of the Pilbara craton have revealed secrets that tell of its very precarious early existence. By about 3.2 billion to 3.3 billion years ago, as Earth was cooling down, a small group of cratons merged into a more coherent, rigid land mass – and so a continent was born.[2] Its remnants still exist in a few scattered places: the Western Australian Pilbara craton, the Western Dharwar and Singhbhum cratons in India and the Kaapvaal

Banded iron formation country, Pilbara region, Western Australia.

craton in southern Africa. This first continent has been named Ur.[3] Sounding like a rather uncertain utterance, this ancient continent was named after the Sumerian city-state of that name dating from 3800 BC. Whereas the geological Ur went on to bigger and better things, evolving into a larger continent and eventually becoming the eastern part of Gondwana, the Sumerian Ur suffered a more calamitous fate in about 2000 BC, being utterly destroyed.

The cratons – the core of Archaean continents – were made of volcanic rocks, attesting to the intensity of volcanic activity on land at this time. (Whether this made it a completely unsuitable habitat for life to develop is a moot point.) Some Archaean microbialitic rocks show indications of having formed in terrestrial hot-spring environments. Is this where life on Earth began? Potential habitats for life certainly also existed in the oceans surrounding these early land masses in a range of environments, from the deep sea to near shore. Continental shelves, important habitats for life in more recent times, were likely to have been restricted around these early volcanic continents. Those areas of shallow water that did exist, however, were in many places colonized by clusters of microbialites. All in all, Earth before 2.5 billion years ago was a very different place from today. Nowhere is this more marked than in the composition of the young Earth's atmosphere, whose secrets are being revealed by rocks.

Step outside in the Archaean and you would be dead within seconds. Studies of minerals that formed on ancient seafloors, along with atmospheric particles that drifted into marine sediments and became incorporated into rocks, as well as rocks that were formerly soils on land, are keys to understanding the composition of Earth's atmosphere. What they show is that it was quite unlike the atmosphere of today. Notably, there was virtually no oxygen in the air, nor was there any dissolved in the oceans. It has been estimated that atmospheric oxygen levels were about 0.000001 per cent of today's level.[4] With very little oxygen in the atmosphere it is likely that there was no protective ozone layer. Life in such an oxygen-free world would have suffered. Dangerous ultraviolet radiation would have affected the upper layers of the oceans, the coastal zone and the land, making it hostile to many forms of life. And to make matters even worse for early organisms, high levels of volcanic activity during the Archaean would have pumped huge quantities of dust and gaseous aerosols into the atmosphere.

Although oxygen may have been missing from the atmosphere, rocks reveal that other gases were present. Nitrogen was a significant component of the atmosphere, with levels similar to what we breath in today, or slightly lower. The other dominant gases in this fledgling atmosphere were methane and carbon dioxide. Methane levels were up to 10,000 times greater than they are today, mainly because there was no oxygen in the atmosphere to destroy it (which otherwise would have been its fate). While we may think that present-day carbon dioxide levels are too high, during the Archaean they were stratospherically higher. On the basis of studies of rocks that represent fossilized soil horizons, it has been suggested that CO_2 levels were between ten and fifty times higher than those at present.[5] Other research indicates it was up to five hundred times higher than present atmospheric levels. Another study has even proposed that CO_2 levels in the Archaean atmosphere were a staggering 2,500 times higher than today.[6]

Another way of assessing the make-up of the atmosphere in the Archaean has been by analysing micrometeorites that were mineralogically altered as they plummeted through Earth's atmosphere 2.7 billion years ago. The alteration products indicate that somewhere between 25 and 50 per cent of the atmosphere was CO_2.[7] Given that the Sun's luminosity was much less at this time, it is probably just as well that greenhouse gases fuelled the atmosphere. Their effects would have provided Earth with a warm blanket, essentially stopping it from becoming a frozen ball spinning in the cosmos and providing an environment conducive to life.

How, then, did this singular atmosphere affect Archaean climate? It used to be thought that early Earth was a much hotter place than it is now. However, recent research points to a climate not too dissimilar to that which we have experienced for the past few thousand years. Surface temperatures, in general, were probably within the range of 0°C to 40°C.[8] There is even a suggestion – based on the discovery in South Africa of Archaean rocks called diamictites, which

Polished slab of banded iron formation, Ord Ranges, Pilbara Region, Western Australia. Alternating iron-rich and silica-rich layers.

are produced by the erosive action of glaciers[9] – of quite extensive periods of glaciation, 2.7 billion, 2.9 billion and 3.5 billion years ago.

Earth's early climate could also have been affected by less-earthly matters. One, somewhat surprisingly, was the speed with which the planet rotated. At present, the Moon recedes from Earth by about 4 centimetres (1½ in.) per year, due to tidal friction of the Moon with Earth. The effect of this has been a steady decline in Earth's rate of spinning since the Archaean, in order to conserve angular momentum. Some rocks contain the history of day length and number of days in the year in the geological past. They show that in the Precambrian both were quite different from today. Annual depositional cycles in banded ironstones from the Weeli Wolli Formation in the Pilbara suggest that 2.4 billion years ago the more rapidly spinning Earth resulted in days being only about seventeen hours long. The consequence of faster planetary rotation rate is to reduce heat transport from the equator to the poles.[10] This would be especially the case if the atmospheric pressure was less in the Archaean. But how on Earth (as it were) to find this out? Well, fossilized 2.7-billion-year-old raindrops, perhaps.[11] What else?

As a star, like our Sun, ages, it generates increasing amounts of light and heat. This is because its inner core becomes denser and the temperature at which hydrogen fuses to helium increases.[12] It has been estimated that 2 to 3 billion years ago the Sun's energy output was perhaps 20–25 per cent less than today's. So, shorter days, more days in the year and a more pallid Sun. As such, this weaker star should have been incapable of stopping the oceans from freezing over. Yet there are rocks showing that liquid water was plentiful at this time – fossil ripple marks, mud cracks and microbialites that only grow in liquid water, all testifying to a watery, unfrozen planet. For this we have to thank the warming greenhouse-gas blanket covering

Earth. While carbon dioxide may be the villain of climate change today, it was Earth's saviour in Archaean times. The presence of liquid water on the planet also means that, like today, it rained a lot.

Although studies of micrometeorites and other rocks show an atmosphere devoid of oxygen and replete with carbon dioxide and methane, what of atmospheric pressure?[13] Seemingly impossible to assess, attempts have been made to extract the secrets of Archaean atmospheric pressure from a most unlikely source: impressions in rock interpreted as fossil raindrops. Though it might seem unlikely that something as transient as a raindrop could be preserved in the rock record, they are actually not that uncommon. To preserve raindrop microcraters made by a passing shower, a wet sandy or ashy surface must be quickly covered by a layer of fine dust or ash. The grain size of this layer must be different from that forming the wet surface, as millions of years later, when the resultant rock splits apart as it weathers, it will do so at this junction between beds of disparate grain size, revealing the fossil raindrop craters.

Among the oldest of what are claimed to be fossil raindrops is a cluster of microcraters discovered in a 2.7-billion-year-old volcanic ash bed in the Kameeldoorns Formation in South Africa.[14] Smaller even than a light impression made by the tip of a little finger, analysis of the geometry of these splattered imprints has, it has been suggested, made it possible to infer the velocities of the raindrops. While many variables come into play, such as the size of the raindrops, the environment through which they fell and the medium in which they landed, by comparing fossil imprints with modern examples and calculating their velocities when hitting a similar substrate, inferences can be made about the velocity of the fossil raindrops at their point of impact. The significance of this is that terminal velocity reached by a raindrop is very dependent on atmospheric pressure. The greater

Possible fossil raindrops on 2.7-billion-year-old volcanic ash bed, Kameeldoorns Formation, South Africa. Passing meerkat for scale.

the pressure, the slower the raindrop will fall. The late Archaean raindrops, it has been maintained, seem to have fallen faster than their modern counterparts as crater size is generally larger. This would indicate an atmosphere 2.7 billion years ago with somewhat less pressure than today, anywhere between 50 and 100 per cent of modern average pressure.

As elegant as this model might seem as a way of deducing atmospheric pressure in the geological past, there has been some criticism of this approach, with alternative suggestions for the causes of the microcraters being proposed, in particular that they represent impressions made by hailstones, rather than rain.[15] The elliptical shape of the fossils could suggest hail impacting the ash bed in a squall. Such an interpretation would mitigate against using these fossils to explain atmospheric pressure during the Archaean.

The fine balance between the amount and the composition of atmospheric gases at this time, and the resultant ambient temperature, had a tremendous influence on creating an environment amenable to the evolution of life on Earth. And it was some of this early life that, in all likelihood, had the transformative power to create great thicknesses of banded ironstones in the late Archaean and early Proterozoic world's oceans. In doing so they also changed Earth's atmosphere irrevocably, and with it the course of the evolution of life and, in turn, the nature of the rocks that encase the planet.

Oxygen Rising

In coastal cliffs, the sides of gorges, road cuttings or quarries, anywhere that the underworld is exposed, rocks often present as distinct, alternating bands, of either different colours, textures or tolerances to the vagaries of erosion and weathering. Take the sea cliffs west of Lyme Regis in Dorset: beds of pale grey limestones alternate over and over again with thinner, more crumbly, darker grey shales. In much the same way, chalk cliffs towering over the English Channel (or La Manche on the French side) often show thick layers of chalk alternating (in an unerringly regular fashion) with distinct bands of narrow, black flint. However, the banded ironstones in the landscapes of places like South Africa, India and Australia are not like this. They are far more complex – banding formed of a hierarchy of alternating rhythms of the different chemical sediments, as sets, both light and dark, and black and red. Layers within layers, like a Russian *matryoshka* doll.

Stand far away from a cliff of banded ironstone. Perhaps 100 metres. The banding will be obvious, with more prominent less eroded layers set about 20 metres or so apart. They dominate the rock face, like a book lying on its side. Its spine hidden from view. The

Banded iron formation, Pilbara Region, Western Australia. Alternating iron-rich and silica-rich layers.

pages, thin layers of chert and ironstone. The story, one of endless repetition played out over scores of millions of years. And hidden within each page there are chapters and within them sentences. Each repeats this same sedimentary paeon of red and grey. Formed from light and dark, heat and cold, they are called macrobands.

Walk towards the outcrop. As you get closer, more layers begin to resolve. Layers just a few centimetres thick, the pages called 'meso-bands', alternating beds of hard, predominantly red, black and grey, repeating time and time again within the macrobands. Move closer still, until you can touch the rock, and within these beds that are often no thicker than your finger, you will see, if well enough preserved, even thinner alternating multicoloured layers. Some repeated bands (the sentences of the book, perhaps) are 'microbands', generally between 0.3 and 1.7 millimetres ($\frac{3}{250}$–$\frac{3}{50}$ in.) thick.[16] Remarkably, within them are yet finer layers a few nanometres thick (1 nanometre is 1,000-millionth of a metre) – the fundamental words of this almost infinitely long story.

One element links the three dominant minerals of which the layers of these banded beds are composed: oxygen. The dark grey to black layers are composed of the iron oxide mineral magnetite (Fe_3O_4). The red layers are also made of iron oxide, but the mineral haematite (Fe_2O_3), while the paler, grey layers are silica in the form of chert (SiO_2). Sometimes the chert is infused with haematite, producing the red chert known as jasper. The centimetre-scale chert layers can comprise up to 60 per cent of the rock, the magnetite bands 20 per cent. The very thin microbands are generally composed of haematite layers, in repetitive bands, with equally thin layers of chert.

It is generally accepted that the banded iron formations were deposited as sediments, although how this took place has intrigued researchers for many years. The principal focus has been on what was so special about the ancient seas 2.8 to 2.4 billion years ago that promoted the global deposition of such vast quantities of iron-rich sediments, often more than a kilometre thick. During the Archaean Eon both the atmosphere and the oceans were essentially free of oxygen. It was a reducing world where the only organisms that existed were single-celled anaerobic microbes. Yet the chemicals that were to be produced worldwide over a few hundred million years are all dependent on oxygen. What was its source?

The oceans in which the sediments that ultimately became banded ironstones and cherts were laid down were thought to have been slightly acidic, because carbon dioxide levels in the atmosphere were much higher than now. As to their salinity, it is likely that Archaean oceans were probably slightly more salty than modern seas. More significantly, like the atmosphere, they were anoxic, lacking dissolved oxygen, but containing huge quantities of dissolved ferrous ions. The combination of iron and silica being produced by hydrothermal activity in the oceans where volcanic material was spewed from vents on the ocean floor, plus iron eroded from volcanic rocks on land and carried into the oceans from rivers, saturated these anoxic oceans with dissolved iron. The absence of oxygen meant that the iron was in its ferrous form, and so soluble in the ocean water.

Whatever life was present on Earth at this time was in aquatic ecosystems in the form of single-celled prokaryotic organisms, probably anaerobic bacteria and archaea, that could thrive in the absence of oxygen. The increase in diversity of microbialites during the late Archaean and early Proterozoic implies an increase in the microbial communities that built them, in particular cyanobacteria. For such organisms that evolved under anoxic conditions and photosynthesized, oxygen was a poisonous waste product to be expelled.

A leading model for how the great deposits of banded iron formations formed around the world in late Archaean and early Proterozoic times proposes that the oxygen generated as a waste product from oxygenic photosynthesis reacted with the ferrous iron dissolved in the oceans.[17] The geological outcome was the periodic precipitation of insoluble ferric oxide – basically, the seas rusted. By approximately 2 billion years ago, just about all of the dissolved ferrous ions in the oceans had been converted into ferric oxide, forming the sedimentary layers of magnetite and haematite. From the analysis of isotopes of trace metals in black shales of this antiquity in northwest Russia, it would appear that both the ocean and atmosphere had become well oxygenated by this time.[18] The oxygen that continued to be produced by the photosynthetic activity of some microbial communities had nowhere to go but up, transforming the atmosphere, utterly, and turning the sky blue.

Astronomical Rock Clocks

Seemingly inconsequential, microscopic bacteria are, then, likely to have played a pivotal role in the precipitation of vast sedimentary iron deposits. Their activity, however, would have been to a large extent subject to the whims of external forces, turning their metabolisms on and off with metronomic regularity: oscillations in temperature, light intensity and nutrient supply. But why do massive outcrops of these rocks show so many hierarchical levels of rhythmic alternating beds – iron-rich sedimentary layers alternating with iron-poor, silica-rich layers from the metre to the nanometre scales? What was driving this cyclicity? What were the conditions that favoured the precipitation of one type of sediment over the other? Or are the rocks we see today just palimpsests of former compositions, the cyclicity developing only as the sediments transformed into rocks? And were the environmental factors responsible for macrobands, mesobands and microbands all alike?

If the precipitation of insoluble ferric oxide was a consequence of microbial activity transforming soluble ferrous ions to insoluble ferric ones by pumping out oxygen, then periods when only silica was precipitated as chert would seem to indicate times when there was little or no microbial activity. Perhaps there were variations in light intensity, any reductions diminishing the extent of photosynthesis. Or maybe, periodically, environmental conditions favoured the activity of iron-reducing, rather than iron-oxidizing, bacteria. Such iron-reducing bacteria are thought to have targeted the iron that was produced on land and transported into the oceans.[19] But this still begs the question of what was controlling the cyclicity.

The different cycles can have separate effects or, when combined, more appreciable impacts on climate variability. Many of these are recorded in the rock record, enabling subtle changes in climate hundreds or even thousands of millions of years ago to be unravelled. It is the impact of many of these cycles, from diurnal to eccentricity, that can be detected in banded iron formations around the world. Here is a potent mechanism to explain the hierarchical nature of their depositional cycles.

The 2.48-billion-year-old Kuruman Iron Formation in Northern Cape province, South Africa, contains well-defined hierarchical patterns of deposition of alternating iron-rich and silica-rich layers, from metres to millimetres. Moreover, groups of some of these coloured bands can be traced laterally for hundreds of kilometres. Clearly, whatever caused the rhythmic change in deposition was controlled by environmental forces that were subjected to global influences. With a sedimentation rate calculated to be in the order of about 10 metres (33 ft) every million years, it has been proposed that the two major sets of cycles, which appear as groups of layers about 5 metres and 20 metres (16 and 65 ft) apart respectively, were both driven by the influence of eccentricity Milanković cycles.[20] One was probably in the order of about 400,000 years, the other in excess of a million years. However, the mystery still remains of how these cycles affected the climate at the time, and thus the type of sediment being deposited. Perhaps variations in extremes of seasonality influenced rainfall, which in turn affected iron and nutrient supplies into the oceans. But then, again, there are some secrets that rocks might prefer to keep close to their chests.

Although now on the opposite side of the Indian Ocean, banded ironstones of much the same age as the Kuruman Iron Formation formed close to those in what is now the Pilbara region of northwest Australia. Described as the world's most extensive banded iron formation, and representing the climax of its deposition in the rock record,[21] the 2.48-billion-year-old Dales Gorge Member of the Brockman Iron Formation varies between 140 and 160 metres (460–525 ft) thick, and consists of seventeen macrobands. Each of these contains centimetre-scale-thick mesobands, which in turn are made up of sub-millimetre-scale microbands of iron- and silica-rich layers. These fine layers are also considered to be the result of astronomical forcing. In this case the rock clock ticked with a yearly rhythm.[22] In the warmer, lighter summer season, iron-rich sediments were laid down, as the activity of photosynthetic bacteria was at its highest. During the darker, cooler winter months their activity diminished and beds of chert formed instead. Remarkably, banding at an even finer scale has been detected, at the nanometre scale. These layers average a mere 26 nanometres in thickness – seemingly little more than the breadth of a shadow. Yi-Liang Li, who made the discovery of these minute layers of haematite granules in the chert, tentatively proposes that these incredibly thin layers could be the result of diurnal precipitation of ferric oxide that arose from the circadian metabolism of ferrous-oxidizing, photosynthetic microorganisms.[23] In other words, the bacteria were active during the day, producing oxygen, but rested by night – a daily rock clock.

Ultimately, these banded ironstones are the product of the Sun. Arguably, photosynthesis was the most important evolutionary innovation in the history of life, with its ability to harness the power of the Sun. As they metabolized, photosynthetic organisms produced oxygen that rusted the oceans and transformed the atmosphere into one in which more complex life forms could evolve. These, in turn, came to influence the nature of many of the rocks that ultimately coated the planet's surface.

Between a Rock and a Hard Place

26 May 2020. It is Sorry Day in Australia. This is an annual day to acknowledge the injustices experienced by members of the 'Stolen Generation' – those Aboriginal children separated (often forcibly) from their families by white Australians. In 2020, this day will also be remembered for the revelation of what was described by Peter Stone, the UNESCO chair in cultural heritage protection and peace, as one of the worst acts of destruction of an archaeological site since the Islamic State destroyed sites such as those at Palmyra in Iraq.[24] The site in question, in the Pilbara region of Western Australia, comprised two Aboriginal rock shelters of great antiquity in banded ironstone formations and were among the oldest documented occupation sites in Australia. Coincidentally, this was also the day I had planned to start writing a simple description of the evidence excavated by archaeologists from this site for the prolonged use of banded ironstones for tens of thousands of years – a site that now no longer exists.

The rock shelters, known as Juukan Gorge 1 and 2, were destroyed by blasting of the banded ironstones by Rio Tinto Corporation to extend their existing Brockman 4 mine. Permission for the blasting had been given by the state government seven years earlier and took place just 11 metres (36 ft) from the shelters. The company is one of the major iron ore companies operating in the region, from which 32 per cent of the world's iron ore (as of 2021) is produced. One of the consequences of the earnest endeavours of the late Archaean and early Proterozoic microbial communities in rusting the oceans was the establishment, in all of the countries where banded ironstones occur, of lucrative mining operations, feeding the world's hunger for steel. The iron ore reserves in Western Australia alone are an astonishing 48 billion tonnes. Much of the mined and exported iron ore is in the form of the iron oxides haematite and goethite. In 2019 China consumed more than 80 per cent of the 878 million tonnes of iron ore mined in the area that year. Its value was almost AU$100 billion.[25]

In 2014, a year after permission had been given to blow up the site, a salvage excavation yielded the oldest bone tool in Australia, the earliest grindstone in Western Australia and a plaited human-hair belt 4,000 years old. However, under the outmoded Western Australian Aboriginal Heritage Act of 1972, such new information does not allow an existing determination to be overturned. Moves are afoot to change this now, although it is too late for the Juukan shelter sites. On the basis of the first excavations carried out in 2008, there was clear evidence that the site had been occupied for at least 33,000 years, with the oldest radiocarbon date being 32,920 years before the present and the youngest 760 years.[26] Somehow this evidence of the antiquity of the site had eluded the decision makers of Rio Tinto, despite it being readily accessible in the archaeological literature. Aboriginal people had been utilizing the rock shelter and the banded iron formation it contained for at least 30,000 years. The 2014 excavation, as yet unpublished, indicates occupancy of the shelters as far back as 46,000 years ago, making it one of the earliest recorded human habitation sites in Australia.[27]

In the preliminary excavations carried out at the site in 2008, 32 stone artefacts were found in one shelter (Juukan-1) and 272 in the other (Juukan-2). Many of these were made of chert or ironstone, both derived from the banded iron formation. The importance of these sites, as well as demonstrating use of ironstones and chert as tools, was the realization that there had been near-continuous occupation of the shelters for over 30,000 years, including during the Last Glacial Maximum that peaked around 21,000 years ago, when the area was particularly cool and dry. The site also demonstrated variations through time in material used for tools, with a preference for ironstone until about 5,000 years ago, when it changed to a predominant use of chert. Why this took place is not clear, but it may relate to climate changes that resulted in changes in hunting behaviour.

A combination of the zealous need to exploit every available outcrop of banded iron formation and a piece of biased legislation has resulted in the loss of a particularly valuable site, to both Aboriginal and archaeological communities. The public outcry following the destruction may, it is hoped, protect other remaining sites, many of which have yet to be studied.

While the use of banded ironstones and chert as tools has a very rich and ancient heritage in the Pilbara region of Western Australia, 46,000 years is almost as yesterday compared with the length of time humans have been using this same type of rock in southern Africa. Situated on a low hill between the Kuruman Hills and the Langberge mountains in Northern Cape province, South Africa, is Kathu Townlands. Here are exposures of banded iron formation, which in the nearby Sishen Mine have been mined extensively as a source of iron ore for more than fifty years. Countless archaeological artefacts litter the ground. Chert dominates the banded iron formation, and as such it has proved ideal for the manufacture of stone tools, a craft that people have been perfecting here for literally hundreds of thousands of years. Archaeological excavations have produced a cache of well-made hand axes that have been dated at about 500,000 years.[28] This is some 300,000 years before the earliest

One-million-year-old Acheulian hand axes from Kathu, Northern Cape province, South Africa, made by *Homo erectus* from chert derived from 2.46-billion-year-old banded ironstones of the Kuruman Iron Formation.

recording of our species, *Homo sapiens*. The makers of these tools may have been members of our ancestral species, *Homo erectus*.

One of the most significant events in human evolution has been the use and control of fire. Just 100 kilometres (62 mi.) east of Kathu, Wonderwerk Cave (*wonderwerk* meaning 'miracle' in Afrikaans) offers significant evidence that banded ironstone brought in by local inhabitants about a million years ago has played its part in humans' early attempts to purposely use fire.[29] The cave itself is formed in a layer of dolomite (calcium magnesium carbonate), but excavations carried out in the cave since 2004 have revealed a layer dated at 1 million years old containing abundant artefacts made from banded ironstone, along with evidence of burnt bones and plant material. Other, much larger pieces of banded ironstone were also found. These slabs were almost certainly brought into the cave, and many show characteristic fracture marks called 'pot-lid fractures'. These usually form when a rock is heated by quite an intensive fire to temperatures in excess of 500°C (930°F).[30] Burnt bone shows evidence of fire with temperatures up to 700°C (1,300°F), which is consistent with a localized fire on the slabs of banded ironstone. The flakes that spalled off the slabs following fire may well have been used to manufacture some of the Acheulian tools found in the deposit.

Thus, for at least a million years, banded ironstone has been a useful resource that could be made into tools to dig, cut, scrape or even kill, and, more recently, turned into washing machines. Yet, when subjected to the prolonged effects of intense tropical weathering, it had a quite different use: art. Around 30 to 40 million years ago, the world was a much warmer place. CO_2 levels were much higher; rainfall was intense. In this humid world, the rocks began to change much faster than they had done for millions of years. In places where water seeped through the rocks, ironstones morphed into other complex iron oxides, including goethite, the brown or yellow hydrated iron oxide. It was also soft, when mixed with clays that were also developing as the rocks weathered, and easily eroded and scraped from the narrow seams in which it had formed. Smeared on dark-red, less-weathered rock, it would have stood out like a streak of yellow light. This was ochre.

One such place where ochre occurs, and has been exploited over a prolonged period, is in Banjima Country in the central Pilbara region of Western Australia. Here, in banded ironstone country, a small quarry, really little more than a scrape in this vast landscape, was worked by the indigenous population to produce ochre to be used for art. Shallow depressions in the hard, banded ironstone that formed the floor of this quarry seem to have been where the ochre was ground into powder. They are polished and show residues of yellow paint.[31] Much of the yellow seam had gone when the site was excavated by archaeologists, suggesting prolonged use of the ochre. While yellow ochre was prized for use in rock art on the predominantly red rocks of the country, few such sites are known close to the quarry. More likely is that the ochre was used for body decoration, and also, perhaps, the adornment of artefacts.

The Banjima quarry is not large. It is set within an overhanging bed of banded ironstone and is just 2 metres wide, 1.75 metres deep and 0.75 metres high (6½ × 5 × 2½ ft). Hardly substantial. But yellow ochre sites were a premium. More common were red ochre sites, also produced from the tropical weathering of banded ironstones. On an altogether grander scale than the Banjima quarry is one of the most extensive and ancient ochre quarries in the country: Wilgie Mia. Meaning 'red ochre place' to the local Wajarri Yamatji people, it lies some 500 kilometres (310 mi.) south of the Banjima yellow ochre mine. Not only is it, by comparison, huge, with an estimated 14,000 cubic metres (18,300 cu. yd) having been removed during the course of its history, but it is extremely old. With a life span of at least 30,000 years, it is the largest ochre mine in Australia, and the oldest continuous mining operation in the world.[32] The longevity of current iron-ore mining operations of banded iron formations throughout the world pales into insignificance.

The mine itself shows a transition from solid, unweathered banded siliceous ironstones to massive haematite, into progressively weathered haematite, and then to the soft, red ochre. The ochre was traded widely through much of Australia due to its quality. Yellow and green ochres also occur in the mine, likewise forming from the deep, prolonged weathering of the ironstone. These multicoloured ochres were not only used for body ornamentation but for rock-art paintings; one site about 60 kilometres (37 mi.) from the mine contains red pigments that match the chemistry of the Wilgie Mia

W. H. Kretchman, 'Natives mining for Wilgie "red ochre" at "Wilgie Mia",
Weld Ranges, Murchison W. A.' 1914.

ochre.[33] To the Wajarri Yamatji people, the red ochre is the blood of
the huge kangaroo Marlu, killed by the evil spirit Mondong, his blood
soaking into the ground. The yellow ochre is said to represent Marlu's
liver, and the green, his gall. Mondong is said to love darkness and
to dwell deep in the mine. The Wajarri Yamatji had a tradition after
finishing working in the mine of walking out backwards, clearing
their footprints from the ground as they went, to prevent Mondong
following and killing them.[34]

* * *

NO LONGER IS BANDED ironstone fashioned into tools to cut, scrape
and kill. Most becomes the steel skeletons of buildings that vie with
each other to become the biggest and the best. The equivalent weight

of tools fashioned from ironstone by early people for thousands of years would now probably be scooped up by huge iron claws from quarries the size of small towns in less than a day. Poured into rail trucks, they are pulled in their hundreds by trains more than 2 kilometres (1¼ mi.) long – modern-day rainbow serpents insinuating themselves across country to the nearest port. The multicoloured loads of ore are sloughed from the trains onto ships that sail to distant lands, to become steel. From this, cities arise. Their ultimate fate, no doubt, will eventually be to fall, to fade, the concrete-coated skeletons destined to return into the self-same rust from which they came, deep in the almost fathomless depths of time.

SIX

RIVERS OF SAND

*Most people's acquaintance with sandstone is somewhat less than salubri-
ous. Being a product of the underworld, it seems to know its place, for it
is often found underfoot in the pavements along which we make our way.
Here it is also often spat upon, shat upon (by dogs, mainly) and generally
downtrodden. We should, perhaps, give it more thanks for the hard-
wearing, resolute nature that arises from its constituent parts – usually a
compact mass of quartz sand, or sometimes pulverized rock fragments,
cemented together by silica. All very resistant to weathering and abrasion.
Who (apart from a few strange geologists), when walking over it, spares
a thought for the ancient landscapes in which it formed: rivers, draining
land largely devoid of animals or shallow seas populated by a menagerie
of creatures long gone from Earth? Occasionally, however, we do find time
to treat sandstone with the respect it really deserves. For sandstone is the
custodian of the behaviour of animals, in particular of the tracks and trails
made by the first animals ever to walk on land. Oh, and we also
crown kings and queens upon it.*

In 1296 King Edward I of England stole a large lump of sandstone from
Scone Abbey in Scotland. The rock was the 152-kilogram (330 lb)
Stone of Scone, otherwise known as the Stone of Destiny or the
Coronation Stone. Edward was not, as far as we know, a closet geol-
ogist. To him it was war booty. For hundreds of years, the rock had
played a central role in the coronation of a succession of Scottish kings.
What better way to insult the vanquished? After being installed by
Edward in a chair in Westminster Abbey, it came to be sat upon by
every subsequent king and queen of England during their coronations.

Stone of Scone – a Devonian-age fluvial sandstone upon which Scottish and British monarchs have been crowned for more than seven hundred years.

But, like most other rocks, it has a far deeper history. It is one that extends back to the time, about 400 million years ago, when the tiny grains of which it is composed were swept together and subsequently cemented to form the sandstone. A detailed geological examination of the stone carried out in 1998 demonstrated that it was derived from a quarry in the Scone Sandstone, part of Devonian-age Lower Old Red Sandstone.[1] But how to read such a rock and determine its early history? A much greater swathe of rocks would help. Even if they are at the other end of the Earth.

A Labyrinth of Stone

It was George Grey's second expedition to northern parts of Western Australia, and given his experience two years earlier, it should have met with some degree of success. After all, his first, in 1837, had been ill-prepared and beset by a boat wreck, a near-drowning, being speared by some less-than-welcoming locals and getting lost, all in the search for a mythical great river. Despite these setbacks he did make invaluable records of the fauna and flora, along with the customs and art of the Australian Aboriginals. This second expedition, in 1839, unfortunately turned into an even worse debacle. George Grey (1812–1898), military man, explorer and later politician and diplomat in Australia, New Zealand and the Cape Colony, had successfully lobbied the Royal Geographical Society in London for support for these expeditions. On 17 February 1839 his party of twelve set sail, heading north from Fremantle on an American whaling ship that carried three boats to be used on his exploration. Their destination was Shark Bay, some 1,000 kilometres (620 mi.) to the north.

After surviving being battered by a passing cyclone shortly after their arrival, they managed to successfully survey the coastline around the bay for several weeks. Their return voyage, however, was even more of a disaster. Following the earlier loss of one boat during the cyclone, they had the misfortune of losing the other two during a misguided attempt to land at Gantheaume Bay, just south of the mouth of a large river that emptied into the Indian Ocean about 600 kilometres (370 mi.) north of Perth. Grey had developed something of a penchant for naming Western Australian rivers that he saw on his expeditions. So, he decided that this one should be called the Murchison River, after one of the most renowned geologists in Britain during the nineteenth century, Roderick Impey Murchison (1792–1871), a founding member of Grey's sponsor, the Royal Geographical Society. Stranded far from Perth with no transport, Grey and his party undertook one of the great feats of early European exploration of Australia: successfully walking back to Perth through land never before visited by Europeans. It took them three weeks, all but one of the party surviving the ordeal.

Beginning their epic southward trek on 2 April, they left the mouth of the Murchison River and crossed 'a red sandstone ravine'. Grey perceptively noted that the rocks were very distinctive in character, 'resembling the old red sandstone of England'.[2] He could equally well have said, 'resembling the old red sandstone of Britain', because the Stone of Scone from the same sandstone unit in Scotland wouldn't have looked too out of place in this terrain. Certainly, such sandstones are geographically very widely dispersed. Serendipitously, just as Grey was staggering across these hot, red rocks, Murchison's great book *The Silurian System* was published in London. The most recent estimate of the age of these rocks, now called the Tumblagooda Sandstone, is Silurian.

The sandstones of the underworld exposed in the steep cliffs of the gorge carved by the Murchison River are, indeed, frequently red. That is when they are not brown, or even yellow, these colours all being part of the story of the life history of these rocks. Should George Grey have chosen to trek the 70 kilometres (43 mi.) or so length of the Murchison River gorge, he would have climbed down through time, descending through a massive 1.3-kilometre (4,200 ft) thickness of sandstone. This is because the strata are gently tilted downwards to the west. John Michell's jaunt back to Yorkshire likewise passed down through such inclined strata. But whereas his journey through time was a colourful expedition, Grey's wouldn't have been. All he would have seen on either side of him in the rugged walls of the gorge for the entire journey would have been sandstone, sandstone and more bloody sandstone. So, where did all this sand come from?

Like living organisms, recycled rocks have different 'life' spans. Some are made in the geological blink of an eye, like the lifespan of a gnat. Others, however, have a maturity that makes the longevity of humans and elephants seem little more than the passing puff of a sluggish breeze. Geologists talk about the sediment maturity of a rock. Some are extraordinarily mature, because the sedimentary grains of which the rock is composed have been recycled over and over again during vast tracts of geological time. In doing so, they are constantly changing in shape and size. For clastic rocks, like the Tumblagooda Sandstone, some of the grains that are now firmly cemented together were eroded from granitic rocks more than a billion years ago. Others, though, were produced 'only' a mere 500 million years ago.

Granitic rocks mainly contain the minerals quartz (silica), feldspar and mica, the constituent minerals in the Stone of Scone. They also have a scattering of rarer minerals, like tourmaline and zircon. Physically and chemically granites are broken down over millions of years by the ravages of weathering, the softer, more soluble minerals dissolving or transforming into other minerals. Feldspars, one of the major components of granites, break down readily into small grains

Poorly sorted sandstone from the 440-million-year-old Tumblagooda Sandstone, Kalbarri, Western Australia. Stained red by haematite; the glassy rounded grains are quartz, the angular cream grains are feldspar.

by the physical action of water or ice, but chemically they will often be changed into clay minerals, like kaolinite (china clay). Quartz crystals, however, the other major mineral component of granites, are barely soluble and are just relentlessly, but very slowly, broken down by physical abrasion. Once broken free from their granitic parents, the quartz grains bounce their way along the beds of streams and rivers and are blown around by the wind. Over extraordinarily long periods of time, the crystal fragments become increasingly spherical and are progressively smoothed and rounded. The closer the grains approach a sphere in shape is one measure of their maturity, and if the rocks have been well sorted, and consist of mostly a single mineral, like quartz, this is another measure of a very mature sediment.

The Stone of Scone, therefore, is not a particularly mature sandstone. Its grains are quite angular, and feldspar and mica are still present.[3] A good part of the Tumblagooda Sandstone, however, is a very mature sandstone, consisting predominantly of rounded, spherical quartz grains, though in places it resembles more the sandstone of the Stone of Scone in having angular chunks of feldspar and rock fragments. With the Tumblagooda Sandstone it is even possible to delve deeper into the rock and, by examining the various rare minerals trapped within it, explore the journey that the individual grains have taken and just where they might have come from.

One mineral found in very small quantities in the Tumblagooda Sandstone, which may hold the key to the birthplace of some of its grains, is tourmaline. Chemically this is one of the most complex of all silicate minerals. Essentially an aluminium silicate, it can contain an incredibly variable range of elements within its structure, including lithium, boron, sodium, aluminium, fluorine, potassium, iron, manganese, vanadium and chromium. In the same way that particular DNA structures are unique to living organisms, many types of tourmaline have a chemical composition unique to particular igneous rocks. This allows their sources to be traced. The detrital tourmalines in the Tumblagooda Sandstone show chemical similarities to samples from the ancient land masses to the east, the Yilgarn and Pilbara regions.[4] Rocks in these areas are very old, many having formed more than 3 billion years ago, so it would seem plausible that this is where all the sand grains that make up the Tumblagooda Sandstone came from. Time to head inland.

A further 200 kilometres (124 mi.) east from where the Murchison River gorge begins are the river's headwaters. Here, gullies incised into a narrow range called the Jack Hills drain into the main river after a passing cyclone has vented its spleen and filled them with rain. Jack Hills rocks are mainly conglomerate – pebbles (indeed any grain more than 2 millimetres in diameter) cemented together. Unlike the mature sandstones through which the Murchison River has carved its sinuous gorge, the Jack Hills conglomerates seemingly lie at the other end of the maturity spectrum. They are, to all intents and purposes, decidedly immature rocks – not much pounding and grinding of the hard quartz pebbles and grains occurred before they were cemented together into a cohesive rock. The pebbles are large, generally the size and shape of a sugared almond. So, poorly rounded, generally poorly sorted and mainly quartz. Not the sort of rock most geologists would get very excited about. But these rocks have, in recent times, attained something of a cult status, and this is all because of what binds the pebbles together. For within the silty matrix tucked in the spaces between them are tiny crystals of the mineral zircon (zirconium silicate).

Rocks like the Jack Hills conglomerates have often led rather unexpected double lives – they are a mixture of the very young (in terms of time spent between parent rock and descendant recycled

rock) and the very old. A strange combination. One part (the immature part) is a mass of pebbles or chunks of rock freshly sourced from
some fast-flowing mountain stream or other highly energetic environment, a beach perhaps. Then there is the binding matrix that holds the
rock together, which might be very different in every respect and had
a totally different, and often prolonged, life history – a truly mature
sediment. How, then, is it possible to unravel these disparate parts?

Imagine an ancient library. A labyrinth of stone. Its stones are
quartz pebbles. Its books – zircons. Each crystal, like any book, is
a repository of information. Locked within it are its age and maybe
where it came from – even other details of the world in which it
was created. The youngest zircons in the Jack Hills conglomerate
are about 3.2 billion years old, which gives a rough age for the stone
library. But there are others that are very much older than this, and
locked with them are also clues to Earth's earliest days. Lurking
within the zircon crystals as they formed deep in the crust in molten
magma were isotopes of uranium. One, uranium-238, decays over
time into the radioactive isotope lead-206. As it is known that the
half-life of the decay, in other words the time taken for half of the
uranium to decay into lead-206, is 4.47 billion years, by measuring
the amount of lead in the crystal it is possible to work out the age at
which the zircon crystal formed.

The fame of the Jack Hills zircons rests mainly with one tiny,
deep purple crystal. In length it is barely as long as the width of
a human hair. Analysis by Simon Wilde and his colleagues from
Curtin University in Perth, Western Australia, showed that the
zircon crystal had crystallized a staggering 4.4 billion years ago.[5] To
put this into perspective, this was merely 150 million years after Earth
itself had coalesced as the Solar System was born. The outer layers
of the crystal had 'younger' ages of about 4.3 billion years, suggesting

This 3.2-billion-year-old conglomerate from Jack Hills, Western Australia, contains the world's oldest mineral, a 4.4-billion-year-old zircon crystal.

that the crystal bounced around in the magma for a while and was recrystallized. This most ancient of all minerals in Earth stone libraries also yielded tantalizing clues to its formation on a nascent planet, showing that, even at that early stage, water existed.

These zircons are all that remain of long-vanished mountains, scooped up and sunken into the stone library of the Jack Hills conglomerate. So, did the zircon crystals follow the tourmalines down a proto-Murchison River, nearly half a billion years ago? Surprisingly not. The zircons in the Tumblagooda Sandstone are much younger, ranging in ages from 1,250 million years to as 'young' as 470 million years.[6] This suggests that the Tumblagooda zircons are not from the nearby ancient rocks that lie to the east and north of the Murchison River. Some, it seems, journeyed from the southwestern part of what is now the continent of Australia. Others, however, seem to have come from much further afield – weathered from rocks that are now in India, Antarctica, even East Africa. This may seem a strange story for the rocks to impart, but when the sandstone was finally cemented together about 470 million years ago, the geography of the world was very different from what it is today. Australia was part of the southern supercontinent which today we call Gondwana, and these now-distant land masses were all snugly attached to Australia. This proto-Murchison River was draining into a rift valley that was being created by present-day Southeast Asia's bid to escape from Gondwana.

The strange story of the grains of sediment in these rocks tells only part of their history. Having established where they came from, in what sort of environment were they deposited? Then how did they transform from slithering sand into a hard rock? Why are some of the sandstones brown, but others red? And why do some, when the Sun's rays hit them, sparkle like diamonds, whereas others, frankly, are merely earthly dull.

On the Move

A lump of homogenous sandstone held in the hand is good at keeping its origin a secret. It will likely be a pretty homogenous pile of sand grains, quartz generally, with little clue as to the environment in which it was deposited. Its colour, though, may give some indication. If it is red, the presence of haematite points to formation in a warm, dry atmosphere; brown or grey indicates a cooler, wetter environment. Yet how did it arrive as a free-moving sediment at the place where it was ultimately cemented and turned into rock? There are two main sources of energy that transport sand grains: wind and water. Each can carry sand grains from just a few metres to hundreds of kilometres. To discover which facilitated their journey it is better to examine more than just a hand-sized specimen. Preferably a large outcrop that exposes as much of the rocks of the underworld as possible. Of all the ancient Palaeozoic sandstones, those lining the Murchison River in Western Australia are some of the best. With cliffs more than 100 metres (330 ft) high and a continuous outcrop of about 70 kilometres (43 mi.) of gently dipping strata, the secret of how these sands were transported prior to deposition and ultimately lithified has a much better chance of being revealed than a single lump of rock. But teasing out this history can be a tricky business, and the interpretation can often be in the eye of the beholder. Scientists, not surprisingly, often don't agree. And such is the case with this spectacular outcrop of sandstone which, in recent years, has been the subject of much geological scrutiny.

The main evidence for whether the sand was transported by wind or water, and if the latter, whether in a river or in the sea, is provided by sedimentary structures. Unlike most other sedimentary rock types, sandstones contain a welter of structures produced by either physical or biological processes. As such they can offer vivid insights into past environments, both the 'lie of the land' and the organisms that inhabited it, including even some of their most intimate behaviours. It's all a case of learning to listen to what the sedimentary structures are trying to tell us.

Although a few sandstones are homogenous masses of cemented sand, most have some sort of bedding. These lines in the rock represent periods when deposition became more sluggish or ceased altogether. When it resumed, the character of the sediment is likely to have been very slightly different, producing a distinctive bed and a distinct lineation – a bedding plane – marking the period of reduced sedimentation. Beds may be little more than a centimetre or two in thickness, or they may be in the order of many metres. Bedding planes in many sandstones are rarely horizontal. Movement of sand, whether it be by wind or water in a unidirectional manner, results in the production of ripples or dunes as the sand is transported over undulating surfaces. As they are transported in a flowing medium, eddying currents excavate already deposited sand and redeposit it down-current as ripples or larger dunes. These are constantly migrating down-current. The steeper slope of a dune or ripple is downstream in the direction in which the ancient current was flowing. When viewed in cross-section in sandstones, the beds are inclined. This is known as 'cross-bedding'. Individual cross-bedded sets can vary in thickness from a few centimetres to many metres. This will depend on the environment of deposition, whether as sand dunes deposited in air or by fast-flowing rivers or tidal flood plains. Bedding plains associated with cross beds can be planar, that is, parallel to the inclined layer above, or they can be curved: trough cross-bedding. These form by the scouring action of fast-flowing water currents.

To travel down the Murchison River gorge is to pass through time, as the gently dipping sandstones tilt towards the setting Sun and the sea. It is to follow the journey of a great river system nearly 450 million years ago, that over thousands of years drained a vast inland Precambrian mountain chain which today has been eroded down to its roots. What the sandstones in the upper reaches of the gorge show is a coarse sand with extensive trough cross-bedding, both indicative of deposition in a series of braided channels, the sands being deposited, scoured and redeposited, time and time again. The troughs themselves are up to half a metre deep and can be followed down-current for many metres. Laterally the channels extend for hundreds of metres. This all points to fast-flowing channels pouring off high mountains onto an outwash plain, sloping westerly to

Planar cross-bedding in Carboniferous Yoredale Series sandstone, Derbyshire.

Trough cross-bedding, Triassic Otter Sandstone, Ladram Bay, Devon.

the sea. What such deposition suggests is that there was little to no vegetation present to inhibit the flow of water and of the sediment. Although the precise age of these sandstones is not known for certain, they are likely to have formed in either late Ordovician or early Silurian times, about 440 million years ago. The only plants that had evolved the ability to survive on land at this time were non-vascular plants, the bryophytes – mosses and liverworts.

The earliest known and undoubted vascular plants were the lycophytes (clubmosses) such as *Baragwanathia* and the rhyniophytes *Salopella* and *Hedeia* found fossilized in Victoria, *Psilophyton* from Libya and *Cooksonia* from the Welsh Borderland, perhaps the most primitive-looking of all plants. All these forms appear in the fossil record in late Silurian to early Devonian rocks, around 420 million years old. Plants like *Cooksonia* contain water-conducting vessels (tracheids) and stomata, firm evidence that the plants were true vascular plants capable of supporting their weight on land.

The presence of fossils of *Baragwanathia* and *Cooksonia* in marine sediments suggests that these plants grew near shore. Before their evolution, land at this time was essentially rock-coloured: greys, blacks and browns. As vascular plants took hold near the water's edge a green mantle soon spread across the world. By the end of the Devonian, forests had covered much of the planet. The effect of this on the nature of river systems was dramatic and saw a transition from early Palaeozoic braided streams transporting and depositing huge quantities of sediments, like the Tumblagooda Sandstone, to the post-Devonian times when plant roots trapped sediment and constrained rivers into more discrete channels. The nature of sediment transport and deposition had forever changed.

* * *

CONTINUING DOWN THE Murchison Gorge, and passing through time, is to experience a change in the nature of the river flow. The coarse trough cross-bedded sands diminish and are replaced by thin sub-parallel sand sheets, often extensively rippled, suggesting deposition of the sands in shallow water under a reduced water flow. These are mature sands, extremely small and well sorted with near-spherical grains of quartz and little else. These sand sheets are interspersed by thin layers of cross-bedded sandstones. But what marks these sandstones as something special is that they carry a cornucopia of biological sedimentary structures – traces made by animals that record their behaviour during some of their first forays onto the land.

On the Run

When animals took their first tentative footsteps on land nearly half a billion years ago, the world was a very different place. There were no meadows. Neither were there forests, for higher plants, from turnips to trees, had yet to evolve. Just mosses and liverworts, clinging precariously to wet rocks. However, into this seemingly hostile and largely barren, windswept world, animals began to venture, forsaking the oceans, lakes and rivers. As the first animals dragged themselves out of their aqueous world onto a harsh land they left the ghostly evidence of their passing in their tracks, trails and burrows. One of the key questions to be answered in order to explain this major step in evolution is which animals came first? How to answer this? First, it needs to be determined whether the tracks and trails found fossilized in rocks were formed by animals walking on land or underwater. There are ways to do this, mostly involving the examination of the intricate details of each footfall, in order to establish whether they formed when the animals walked with the sun on their backs or were clothed with an aqueous cloak.

Evidence from fossilized tracks preserved in rocks, particularly sandstones, suggests that the first animals to take this step were arthropods. Although beautifully adapted to a life on land, with armoured, jointed legs capable of supporting the weight of their hard, shield-like bodies, whether these intrepid creatures were part of a fledgling terrestrial ecosystem or were just chance adventurers venturing onto an alien landscape is not always clear. In all likelihood they were probably opportunistic terrestrial travellers, migrating from pool to pool on exposed sandflats that formed in either estuarine or river environments, but generally more at home in an aquatic environment.

To stand on some of these sandstone surfaces where the land's first animals walked is the closest we can ever get to travelling back hundreds of millions of years in time. It is possible to walk along the edge of what were once shallow pools of water preserved for about 440 million years. Some show strand lines preserved along their margins as the water level sank, and parallel to these strand lines are the footprints left by multi-legged arthropods that wandered along the water's edge, perhaps feeding from organic detritus left by the subsiding water. Other surfaces preserve the tracks emerging out of the water onto exposed sand flats from depressions that once were shallow pools.

One surface reveals the intimate behaviour of an arthropod as it returned to the water. A sloping surface that was once the edge of a stream or tidal channel. In the deeper water, distinct ripples. On its margin where the sand is wet, a microbial mat has formed, giving the sand a cauliflower-like texture. The water gets deeper as a gentle slope passes down into the faster-flowing water body and the rippled sand. At the water's edge are faint traces of where an

The 440-million-year-old Tumblagooda Sandstone, Murchison River, Kalbarri, Western Australia. In foreground, trackway about 30 cm (12 in.) across made by an arthropod, possibly a eurypterid.

arthropod has walked. As it began to move from land to channel its belly dragged on the soft microbial mat, displacing a piece about the size of a hand and pushing it into the water. As it became deeper the coherent mat dropped down onto the gently sloping sand. The animal had essentially surfed on the mat into the water. The last push of the arthropod before it moved into the deeper water is also miraculously preserved. In what are now hard sandstones is preserved, seemingly for eternity, one animal's action for about ten seconds of its life 440 million years ago.

What, then, of this surfing animal? The paired rows of tiny foot imprints it left behind show that it was, almost certainly, one of a number of different types of arthropods whose tracks are preserved in these fine, well-sorted sandstones in the Murchison River gorge. The different types of arthropod tracks are characterized by the number of repeated sequences of imprints made, reflecting the number of walking legs possessed by the animal, and by the width between the parallel sets of tracks. This provides an indication of the animal's size. Four main types of tracks have been characterized in these particular sandstones. The surfing arthropod left rows of imprints spaced about the width of a small hand apart. Other similar, better-preserved trackways have been discovered revealing that they were made by an animal that scuttled along on eleven pairs of walking legs. Fortunately, this corresponds with the only known body fossil of an arthropod found in these sandstones after about forty years of searching: a euthycarcinoid called *Kalbarria brimmellae*.

Looking like an elongated cockroach with more pairs of legs, *Kalbarria* provides a clue to how this arthropod, and the many trackways, came to be preserved in this fine-grained sandstone. The body fossil is an impression of the underside of the animal. It is as though it had been forced down onto a wet sand surface to make an imprint – think pushing a cockroach down onto some plasticine. In the case of this *Kalbarria*, its demise likely followed being caught in a sandstorm while walking along a wet, exposed sand surface, ultimately becoming smothered in fine, dry sand. The slight difference of grain size between the wet sand surface and the windblown sand meant that when cemented into rock this was a line of weakness along which it subaerially weathered, exposing the euthycarcinoid some 440 million years after it was buried.

Many of the trackways, when not moving in and out of what were once pools of water, just show animals walking, seemingly purposefully, in pretty much a straight line. But there is more than one example when the euthycarcinoids left tracks exhibiting some rather strange behaviour. The tracks they left behind in the sand are arranged in almost perfect spirals. At first glance what appears to be a track progressing in a simple straight line begins to make a wide curve to its left and progressively spirals two and a half times before stopping. The width of the spiral is about half a metre (20 in.). Could this have been the death throes of an animal left stranded out of water? However, closer examination reveals that the animal was not making a tighter and tighter spiral. It was actually walking in the opposite direction in an ever-widening spiral, before progressing onto the straight and narrow. Getting into the mind of this animal and its bizarre activity is even beyond the imagination of an alleged intelligent primate living some 440 million years in the future.

Other trackways preserved on the sand sheets were made by very small animals, barely an index finger in width, the smallest just 5 millimetres (⅕ in.) across. Yet where they tracked across sand ripples it is clear that in some instances the imprints of their appendages were only preserved on the damp crests that must have been exposed to the air, not in the swales in which water was still puddling – more evidence that these animals were taking tentative footsteps on land. Even the largest of the trackways made by arthropods carry signals for this incipient terrestriality. With an outstretched hand's width between rows of the footprints made by the appendages, the exquisite preservation of some individual footprints is such that they could

Tumblagooda Sandstone, 440 million years old, Murchison River, Kalbarri, Western Australia. Ripple-marked surface across which walked a synziphosurid arthropod. On the right is a smaller trackway made by a euthycarcinoid arthropod.

only have been formed and preserved on wet sand exposed to the air. The animals were land-borne.

These larger trackways show repeated sets of eight pairs of footprints. Eight legs, along with some impressions left behind as they rested in wet sand, reveal the animals to belong to a group called synziphosurids. These were ancestors of the modern king 'crab' *Limulus*. Rarely, even larger sets of tracks occur, some up to about 30 centimetres (12 in.) in width. Their stride pattern suggests animals walking on three pairs of legs, probably eurypterids, or 'sea scorpions'. Growing up to 2 metres (6½ ft) in length, these top-ranking predators were also just part of this transient amphibious ecosystem, made up predominantly of predators, whose presence at this time has been preserved in often exquisite detail in these fluvial and tidal sands.

The preservation of these footprints, in such fine detail, with impressions made in sand retaining vertical walls, can only have occurred when the sand was wet, exposed to the air and not submerged; think sandcastles. Research on wet granular piles (that is, sandcastles) has shown that on wet sand surfaces that are subaerially exposed, surface tension between sand grains will provide a rigid, cohesive force over a wide range of water saturation. It is only in extremely wet or extremely dry sands that footprints will be ill-defined; in wet, cohesive sand each individual footprint can be preserved in exquisite detail, the walls of the footprints often being steep-sided to vertical because the wet sand was sufficiently cohesive when the foot was withdrawn from the sediment to retain its exact shape. Moreover, the animal's direction of movement can be interpreted from small, discrete mounds of sediment piled up behind each footprint – the first sandcastles.

It is likely that the forays of these arthropods onto *terra firma*, so exquisitely preserved in these sandstones, were short-lived, as the animals moved between pools of water that were drying out. Whether these sand plains covered by shallow ponds and channels were part of a flooding river system or a tidal sandflat has been a matter of much scientific contention in recent years. One clue lies with the Moon. Around the time of the deposition of these sands about 440 million years ago, it was much closer to Earth. Consequently, tides would have been more extreme than today. An examination of fossil corals 370 million years old carried out in the 1960s showed that there were about four hundred daily growth rings formed each year.[7] This implies that at that time, each day was only 21.9 hours long, because Earth at that time was spinning faster. Since the formation of the

Moon two things have happened: it has steadily moved away from Earth, and as a consequence its effect on Earth has been to slow the rate at which Earth spins. This is due to tidal friction caused by the tidal bulge that the Moon produces as it rotates around Earth. Like Earth, the Moon rotates anti-clockwise (that is, when viewed from the North Pole), but more slowly than Earth. Thus the point on Earth where the tide is highest was underneath the Moon a short time earlier. High and low marine tides are slightly delayed by tidal friction, largely due to turbulence and drag in the oceans. As the tidal bulge is largely caused by the Moon, it tries to pull the bulge back into alignment. This imposes a torque on Earth, causing its rotation to slow down. Astronomical observations taken over the last two centuries have shown that tidal friction is lengthening the day by about one-fifth of a second each year. This is corroborated by the evidence from fossils preserved in rock.

The steady movement of the Moon away from Earth is occurring at about the same rate that our fingernails grow. Ocean tides would have been, therefore, much greater 450 million years ago, and it is hard to comprehend how tides that raised the sea level daily by tens of metres and rushed kilometres up river channels would have preserved the delicate trackways found in the Tumblagooda Sandstone. Given the evidence from the fossil record that there was little in the way of terrestrial vegetation at this time, reworking of the sand pouring down into the near-shore river systems as windblown sands would have not been infrequent.

On the Hunt

If sandstones are capable of preserving delicate footprints reflecting the comings and goings of long-extinct arthropods, it would not be surprising to learn that these rocks can also preserve other, somewhat more violent behavioural traits. When not moving from one body of water to another, much of the time spent by these amphibious arthropods would have centred on hunting for food. Evidence for hunting behaviour fossilized in sandstones comes from the association of particular types of burrows made by what are presumed to be predator and prey.[8]

The most extensive burrows are a type that are called *Heimdallia*. They are widespread in rocks of this age, being found in sandstones as far apart as Antarctica and Ireland. They are complex burrows with both a vertical and horizontal component, occurring in the horizontally bedded sand sheets. Often these beds are wave-rippled, indicating that they formed in shallow water. In plan view the burrows vary from straight to extremely sinuous, and between 5 and 20 millimetres ($\frac{1}{5}$ –$\frac{4}{5}$ in.) in width. The burrows can occur in large numbers, resulting in beds of sandstone severely reworked and looking not unlike beds of fossilized macaroni many square metres in area and up to 12 centimetres (5 in.) in thickness.[9]

Heimdallia burrows found in Devonian sandstones in Antarctica are thought to have been made by arthropods essentially mining the sand and extracting organic material.[10] The extent of *Heimdallia*-rich beds in the Tumblagooda Sandstone implies that periodically the shallow bodies of water in which they formed were organically very rich, possibly in algae or bacteria, or both. This may explain why the sand would have been sufficiently cohesive to preserve the burrows. Some 350-million-year-old Carboniferous examples from Ireland have also been interpreted as feeding structures made by arthropods slowly moving through the sand.[11] It is likely that the narrow, finger-width tracks were made by this self-same *Heimdallia* animal when it made occasional forays onto land.

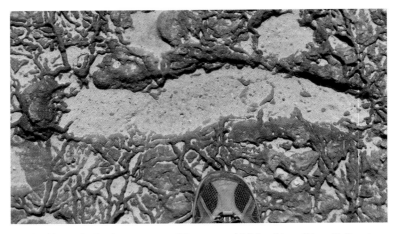

Tumblagooda Sandstone, 440 million years old, Murchison River, Kalbarri, Western Australia. Large hunting burrow made by synziphosurid arthropod, entering from the right, hunting small animals that made the many meandering burrows.

These heavily burrowed sandstone beds are often associated with much larger, horizontal to sub-horizontal burrows that are anywhere between 4.5 and 8 centimetres (1⁴/₅–3 in.) wide and up to 40 centimetres (16 in.) long, although most are only about half of this. The burrows seem to have been made by arthropods making 'scoops' into a wet sand surface, pushing the sand aside or upwards. Sometimes the sand has been pushed anteriorly and forms a crescentic mound at the front of the burrow, providing evidence for the direction of movement of the burrow creator. These burrows are common in *Heimdallia*-rich beds and could represent hunting traces made by the arthropod, feeding on the organism responsible for making the *Heimdallia* burrows. Given their size, there is a reasonable chance that the predators could have been euthycarcinoids.

These same animals may also have been responsible for the production of other traces where they used a different hunting technique when pursuing another type of prey. On some sandstone bedding surfaces are preserved crescentic hollows, deeper at one end than the other. Most, like the horizontal hunting burrows, are 4.5 to 8 centimetres in width, reaching up to an anterior depth of 2 centimetres (⁴⁄₅ in.). The anterior margin may be either inclined or vertical, or even sometimes overhanging.[12] The lateral margins may sometimes show transverse scratch marks made by the appendages as the animal was digging in the sand. Single surfaces can have many dozens of these imprints, often aligned in linear groups. These probably represent the activity of a single animal, repeatedly digging in the sand at low to high angles.

As to what it was and what it was hunting for, the shape of the crescentic impressions, combined with their size, suggests that again the excavator was a euthycarcinoid. Its prey lived in very small U-shaped burrows. When viewed from above, paired holes, each only a few millimetres across, were the external expression of vertical tubes that were connected about 3 centimetres (1 in.) down in the sand. The occupants were likely to have been either worms or tiny arthropods, and the prey for the predatory euthycarcinoid.

Sands deposited in such fluvial or tidal settings, where short periods of environmental stability allowed organisms to leave impressions of their behaviour in the sediment, combined with suitable conditions for their preservation, make sandstones ideal environments for revealing details of trophic structures and behaviours of long-extinct organisms. And many outcrops of these early Palaeozoic sandstones are common worldwide, thanks, it has been argued, to the lack of plants on land at the time.

* * *

WHAT OF THE STONE OF SCONE? It is not just a random lump of sandstone. Geological analysis revealed that, like the older parts of the Tumblagooda Sandstone, it consists of trough cross-bedded sands, each bed 1 to 3 centimetres thick and composed of feldspar-rich sands that grade up from coarse to fine, reflecting a decline in the energy of the flow of water that transported the sediment in each fluvial pulse.[13] The seat of monarchs began its life about 400 million years ago, in an alluvial sand plain at the base of a mountain range that would have been largely bare of vegetation. Across it meandered braided streams, transporting sand that ultimately cemented into the Stone of Scone: a sandstone fit to be perched on by kings and queens.

SEVEN

TURN OF THE SEASONS

*Rocks. We just take them for granted. They are all around us – well,
beneath us at least – and ripe for exploitation. They are the bedrock upon
which life on this planet has flourished. But whoever stops to give rocks
a second thought? To appreciate them? To get inside them to understand
what they are or where they come from – to delve into their innermost
secrets? Yet even the most seemingly insignificant lump of rock often has
much to tell us about the history of this planet: of changing environments,
of the pulse of life. And some are the keepers of the secrets of past climates.
Rocks have been worshipped by people and even used to protect their
monarchs, but also tell tales of the cycles of the seasons and how
even they had the power to create rocks.*

Sarsens, Puddingstones and Silcretes

West Woods sits high on chalk downland in Wiltshire. Three
kilometres (nearly 2 mi.) to the northeast lies the Neolithic
stone circle at Avebury. Twenty-five kilometres (15½ mi.) to the south,
Stonehenge dominates the landscape. Now a small beech wood carpeted by bluebells in the spring, West Woods was a very different
place 4,500 years ago. People lived and died here then, farming the
land and leaving large burial mounds to honour their dead. Some
also spent time quarrying rocks – but not the chalk that forms the
land in this part of the world. It was quite a different sort of rock,
one that capped the hills, or sometimes slumbered in the valleys. A
rock perfect for building megalithic monuments, like Stonehenge
and Avebury.

Neither white nor soft, this rock that so interested Neolithic people was grey and very hard and littered the landscape, resting on the chalk, some say, like large flocks of sleeping sheep. Known from at least Saxon times as sarsens, these rocks range from fine to coarse sandstones. Usually grey, some may be brown, caused by the presence of iron oxides. 'Sarsen' is a word thought to be Anglo-Saxon in origin, derived from either *sar stan*, a 'troublesome stone', or *set stan*, a 'great stone'.[1] What all sarsens have in common, however, is their tough silica cement, which often makes up most of the rock, hence their geological name: silcrete.

Silcrete differs from other types of rocks in two fundamental ways. First, it is not a sediment formed in the sea, nor in a lake or river. Second, it is characterized not by its detrital elements, like repurposed sand grains or broken shells, but by the nature of the cement that binds these particles together. For this reason, silcrete can come in a multiplicity of guises, from the finest of sands through to the coarsest of pebbly conglomerates. But all share the common feature of having the clastic grains cemented together by an extremely strong silica cement. Silcrete is as hard as flint. As hard as chert. While the sands and pebbles from which silcrete is formed have often been derived from river or beach deposits, it can also develop on land from the broken, rotted debris derived from deeply weathered, igneous rocks, like granites. Of the three basic component parts of granites – quartz, feldspars and micas – the latter two are subject to the ravages of chemical weathering much more than the quartz. Thus much of the residual material from deeply weathered granites is quartz, although this itself will slowly be subjected to dissolution under conditions of high temperatures and high rainfall. Some types of silcretes can even consist of more cement than anything else.

Stonehenge, Wiltshire. Outer uprights and lintels made of silcrete sourced locally from West Woods.

Because different silcretes can have very different sedimentary histories, hence quite specific chemistries, it is possible to identify the original locations of those used for building purposes. For instance, detailed geochemical analyses of the Stonehenge sarsens, along with a large suite from twenty sites across southern England,[2] have revealed that of the 52 remaining sarsens from the original approximately eighty stones used to build Stonehenge, all, bar two, came from a single site: West Woods. Given that they appear to have been sourced from this one place, it strengthens the view that the erection of all the sarsens at Stonehenge occurred at much the same time, about 4,500 years ago. But why from West Woods, given its distance from Stonehenge? The most likely explanation is that it was simply a question of the size and quality of the stones. However, given that some weigh up to 30 tonnes, how they were transported the 25 kilometres (15½ mi.) to Stonehenge remains a mystery.

Sarsens occur widely across southern England, from Suffolk to Devon – basically wherever there is chalk. Their greatest concentration is in Wiltshire, particularly the Marlborough Downs.[3] Over the millennia, many sarsens have been removed from chalk downland to be used as building stones – in walls, houses and even a castle. When Henry II ordered Windsor Castle to be reconstructed in stone in the twelfth century, sarsens were used extensively, such as in the Round Tower and Norman Gate. These sarsens were sourced from Bagshot in Surrey.[4] Despite the gradual denudation of the fields of sarsens over the centuries, we know that they were still a prominent feature of the landscape in Wiltshire in the seventeenth century, thanks to one man's diary. A man who was fighting for his king.

It is 12 November 1644, and Richard Symonds, along with about 7,000 others in King Charles I's army, is marching west, following a less-than-successful engagement with Oliver Cromwell's Parliamentarian forces. It is 'a miserable wett, windy day' as the army moves 'over the playnes to Marlingsborough, where the King [lies] at the Lord Seymour's howse; the troopes to Fyfield, two myles distant'.[5] This location, long renowned for the quantity and size of its sarsens, impresses Symonds. It is 'a place so full of a grey pibble stone of great bignes as is not usually seene; they breake them and build their howses of them and walls, laying mosse betweene, the inhabitants calling them Saracens' stones'. This, in all likelihood, is how Symonds probably interpreted the locals' pronunciation of 'sarsens'. But it is the extensive spread of the sarsens that so impressed Symonds, for he recorded how 'in this parish, a myle and halfe in length, they lye so thick as you may goe upon them all the way. They call that place Grey-weathers, because a far off they looke like a flock of sheepe' (a wether being a castrated lamb).[6]

Boulder of 'Hertfordshire Puddingstone' about 60 cm (24 in.) across, from
Collier's End, Hertfordshire.

Sarsens are not the only silcretes to litter chalk downland. The
silicification event that formed the sarsens certainly affected the
detrital sands deposited on the chalk, but it also targeted pebbly
gravels. These coarser sediments are usually flint pebbles derived
from erosion of the underlying chalk. Cemented together by silica,
they form a hard conglomeratic rock known colloquially as 'puddingstone' (shortened from 'plum-puddingstone'),[7] the most distinctive
being 'Hertfordshire puddingstone'. Spread across much of southern
East Anglia, the Chilterns and Normandy (where they are known as
poudingue),[8] they may well represent 'a shore-deposit . . . the shingle-
bed of flint-pebbles consolidated by the infiltration of silica'.[9]
Puddingstones have intrigued people for a long time. In the late
Iron Age and Roman periods (1st century BC to 1st century AD) they
proved ideal to be used as querns – hand mills for grinding flour.[10]
More than seven hundred Hertfordshire puddingstone querns have

been found all over East Anglia and Normandy. This is likely to have largely denuded much of the Chilterns of boulders and concretions of the puddingstone. It has been estimated that the Romano-British quern industry would have used more than 560 tonnes of the rock in a period of less than a hundred years.[11]

By analysing the nature of the pebbles, the matrix in which they are set and the silica cement that binds them together, it is possible to lay out the history of how these curious rocks were formed. First, the sea levels began to recede in early Palaeocene times, about 60 million years ago, following their high stand during the Cretaceous. Sea levels then rose again, rounding angular flints to form rounded pebbles. As the sea level continued to rise, pebbles and fresh flints were mixed together. The sea then receded once more, exposing a beach of rounded pebbles. Exposure to the air resulted in some becoming stained red by a coating of the iron oxide haematite. The pebbles were mixed with a matrix of sand, and this became cemented with silica as groundwater that had been suffused through the sand and pebble mix evaporated.[12]

What the different forms of silcrete have in common, be they sarsens or puddingstones, is that the precipitated silica which forms the essence of the rock was derived from groundwater that has been subjected to strongly fluctuating water table conditions. Prior to developing as silcrete, the porous, unconsolidated clastic sediments that ultimately cemented together needed to have undergone regular wet and dry cycles. Water supersaturated with silica saturated the sediments during wet periods, when the water table was high. In the dry season, the water table fell and the silica crystallized, binding the grains together. The process is not unlike the heinous crime of placing a wet teaspoon in a bowl of sugar and leaving it overnight, the dissolved sugar recrystallizing and binding the sugar grains together on the teaspoon. The wet and dry periods that control silcrete's formation are, therefore, driven by climatic cycles, and are typical of environments that have a highly seasonal climate – extremes in both seasonal rainfall and temperature.[13]

For silica to be dissolved and concentrated in groundwater, ambient temperatures need to be very high, as silica becomes more soluble as temperatures get higher. In the case of the sarsen stones of Stonehenge and Avebury, and much of the fabric of Windsor Castle, they are considered to have formed during a period of Earth history between about 47 and 56 million years ago – that is, during the middle Palaeogene, more specifically early Eocene, times – when global temperatures were extremely high. (The Palaeogene comprises the Palaeocene, Eocene and Oligocene Epochs and lasted from 65 to 23 million years ago.) This event, which saw a rapid and profound elevation of global temperatures, is known as the 'Palaeocene–Eocene Thermal Maximum' (PETM) and began about 56 million years ago. For the following million years or so, Earth was a hothouse world, with average global temperatures more than 10°c higher than at present.[14] By contrast, today (believe it or not) we live in an icehouse world.

What caused the great spike in temperatures at the Palaeocene–Eocene boundary has been the subject of much debate in the scientific literature. Currently the most favoured explanation centres on the effect of the development of what is known as the North Atlantic Igneous Province. Extensive volcanic activity, thought to have been caused by a pulse in Earth's mantle convection about 56 million years ago, beneath where present-day Iceland was then situated, caused Palaeogene Scotland to be raised above the waves.[15] At the same time, southern Britain was being squeezed from the south as the African plate moved north into the Eurasian plate, pushing the chalk sediment up out of the sea. While there was extensive erosion of the chalk, much persisted in southern Britain. Upon this new land that was to become the British Isles of today, sand and gravels accumulated, derived from the weathering of the northern and western land masses – present-day Wales, the Lake District and Scotland. These had once been scattered islands in the vast chalk sea. As a consequence of the volcanic activity in the North Atlantic, huge volumes of greenhouse gases, both methane and carbon dioxide, belched into the atmosphere. The result: the abrupt rise in global atmospheric temperatures. This led to an intensification of chemical weathering of rocks, leaching silica into the groundwaters, and so began the silicification of the sediments that carpeted the chalk lands, ultimately forming the rocks that enabled Neolithic people to construct Stonehenge.

The environmental factors responsible for the formation of silcretes during Palaeogene times were global in their effects. This is apparent because not only do silcretes occur extensively across much of southeastern England and northeastern France, in the form of sarsens and puddingstones, but they are found just as abundantly on the other side of the world, particularly across much of Australia. In southwestern Western Australia, the intense chemical weathering that occurred as a result of the increased volcanism associated with Australia rifting away from Antarctica 43–38 million years ago, belching increased amounts of CO_2 into the atmosphere, is reflected in both the formation of silcretes, now capping many hills made of Precambrian rocks, and the formation of extensive deposits in marine environments of another siliceous rock type known as a spongolite.[16] As its name implies, this rock, light both in colour and in weight, formed from an accumulation of the siliceous spicules derived from dead sponges that thrived in estuaries and embayments off the southern coast of Western Australia, as large amounts of silica-enriched groundwater poured into the sea.[17] The land and the oceans were

Tropically deeply weathered granitic and gneissic rocks, now preserved as
saprolite (yellow), kaolinite (white) and ferricrete (brown), Corackerup Creek,
Boxwood Hill, Western Australia.

awash in silica. Such marine silica enrichment was not confined to southern Australia. Eocene times are known to have been an interval of greatly increased silica accumulation in the oceans globally. This arose from the overall deep weathering of terrestrial environments in high latitudes, and accumulation of the released silica into sluggish oceans. By the end of the Eocene, about 35 million years ago, there was sharp global cooling, stimulating oceanic upwelling and biosilicification. Shallow seas became forested by sponges.

On the ancient Precambrian land surface of the Yilgarn Block of Western Australia, the strongly silicified rocks form thin drapes on hilltops and developed from extreme chemical weathering of granitic rocks, particularly feldspathic minerals and micas, which yielded much of the silica that found its way into the groundwater. Such weathering during the Eocene hothouse world was a very complex and prolonged process, resulting in an upper weathered horizon, a few metres thick, of sand-saprolite and a lower horizon of the clay kaolinite.[18] Saprolite is essentially a rotten rock (from the Greek word *sapros*, meaning 'rotten' or 'putrid') and is a name given to partially chemically weathered granitic rocks. Acidic conditions were necessary to cause such weathering; the deep weathering and constant passage of water containing dissolved silica is further evidence for a warm, humid climate at this time. Associated thick layers of kaolin clay also strongly indicate formation under warm and humid conditions. Silcretes come in two types: one that forms deep in the weathered zone, as so-called groundwater silcretes, and another that develops in the upper soil horizon, known as 'pedogenic silcretes'. The sarsens in southern England are thought to be of this latter type. Evidence comes from the presence of hollow tubes often running through the rock – the former courses of plant roots. In Western Australia pedogenic silcretes develop close to the top of the weathered horizon, on top of the saprolite, further emphasizing periods of high water table. To obtain sufficiently high concentrations of silicic acid, allowing subsequent precipitation of silica, it is crucial for there to be a significant reduction in the passage of groundwater flow. An appreciable increase in the extent of evaporation is therefore necessary. This can be due to either change to somewhat drier conditions or the establishment of a more distinct seasonal climate.

In such a seasonal climate there would have been periods of dissolution when warm and humid conditions alternated with episodes of silica mineralization during the drier season. The temperature must still have been sufficiently high to cause evaporation.[19] While it is possible to infer a strongly seasonal climate during Eocene times in southwestern Australia, the climate was very different from the present day, when during the hot summer months temperatures are often greater than 40°c, rainfall may be non-existent and humidity extremely low. During the Eocene, southwestern Australia was located at a latitude of about 55°s, compared with a little under 35°s today. As a result, day length would have been long in the summer, but very short in the winter. Combined with a wet season that was also very hot, and a dry season that was still very warm, conditions were quite unlike anything today.

Under such a climatic regime, it is interesting to consider how plants would have coped. Fortunately, many of the plants that were growing in the sands before they turned into silcrete dropped their leaves, fruit and seeds, which left their impressions in the silcrete. Thus the fossilized remains of the plants that lived just where the silcretes were forming are all preserved. Today, deep patches of sand are present on the ancient Precambrian granitic rocks, testament to a prolonged period of deep weathering. Such deep sands must have likewise been present during the Eocene. In these, a rich flora flourished. Analysis of the floral communities preserved in the silcretes has made it possible to gain an added insight into the climatic conditions under which the rocks formed.

Seasonality increases toward the poles because of the greater extremes in day length, with much longer summer days and correspondingly shorter winter days. As a result, the flora would have experienced overall darker, though not necessarily colder, winters. The ambient climatic conditions would have been substantially warmer than in equivalent modern southern latitudes. Many of the plants consequently possessed scleromorphic, long-lived leaves. These are thick, closely spaced, short leaves that are adapted to low nutrient levels and low phosphorus in the soil, but are also an alternative strategy to deciduousness, allowing plants to survive the low-productivity dark season irrespective of water availability. Today such leaf architecture is typical of kwongan vegetation in southwest Australia, the chaparral in California and the maquis of the Mediterranean region. As the continent dried out in post-Palaeogene times, such a leaf type was a preadaptation to xeromorphy (low water availability).[20]

The fossil plants preserved in the southwestern Australian Eocene silcretes show a fascinating mixture of forms that persist into today's floras, having adapted to a Mediterranean-type climate of hot, dry summers and cool, wet winters living alongside rainforest plants. These are now absent from the region today, but flourish in rainforests in eastern Australia and elsewhere, such as southern Chile and New Caledonia. Despite growing in high latitudes, the fossils of leaves and fruit from rainforest trees provide evidence of a much warmer and wetter environment in the Eocene when the silcretes formed.

Notable fossils to occur in the silcretes, now extinct in the southwest of Australia, but which are found in modern rainforests, include

Silcrete, 35 million years old, from Walebing, Western Australia, preserving floral elements still extant today: proteaceous leaves (left and right), along with extinct rainforest elements, *Nothofagus* leaves (centre).

the southern Antarctic beech (*Nothofagus*) – extant in eastern Australia, Tasmania and Chile; the she-oak, *Gymnostoma*; the Kauri pine, *Agathis*; the podocarp conifers *Dacrydium* and *Phyllocladus*; *Macadamia*; and other as-yet-unidentified broad-leafed forms with typical rainforest leaf architecture. The scleromorphic plants growing alongside the rainforest trees in the Eocene included many proteaceous genera, such as *Banksia* and *Grevillea*, that have diversified and flourished in the southwest today.

* * *

AT FIRST THE GOING was slow, George Grey and his men battling 'thick scrub, through which we could only make our way with great difficulty', as they began their epic 600-kilometre (373 mi.) journey on foot southwards to Perth in Western Australia in April 1839. It was late afternoon when they reached a small ravine and 'were then able to move along with tolerable facility'. This was where Grey observed that it resembled 'the old red sandstone of England' and he and his men spent the evening 'clambering about the rocks and endeavouring to avoid such natural obstacles as impeded our route'. As night began to fall, he made his way out of the ravine with Frederick Smith to higher ground,

> and on climbing to its summit it appeared to be so well adapted for a halting-place for the night that I determined to remain here. A wild woodland and rocky scenery was around us; and when the moon rose and shed her pale light over all I sat ... gazing alternately into the dim woody abyss below, and at the red fires and picturesque groups of men, than which fancy could scarcely image a wilder scene.

They had clambered over the Silurian Tumblagooda Sandstone and, unwittingly, spent the night on a hard layer of silcrete which covers the older rocks like a blanket. This silcrete, like the Eocene silcretes of Stonehenge and other parts of southwest Australia that had formed from sands into which plants had insinuated their roots, and onto which leaves and fruit fell, had also once been sand – sands weathered from the underlying Tumblagooda Sandstone and heavily silicified about 100 million years earlier. But the fossil plants that lay in the bed of rocks on which the two men slept were quite different. They had grown at an altogether earlier time in Earth history, a time even before flowering plants had evolved. The silcrete Grey and Smith had slept on was an early Cretaceous land surface, one over which dinosaurs had once walked. The evidence for the age of the silcrete comes from the fossilized remains of plants preserved in the silicified sands. Some of the fossil plant species are found much further north in the state, in the Broome Sandstone, a unit dated as early Cretaceous in age. Nearly half of the plant species in the silcretes upon which Grey and Smith slept are ferns – royal ferns (osmundaceans), forked ferns (gleicheniaceans) and umbrella ferns (dipteridaceans); a quarter are cycad-like seed ferns (bennettitaleans), now long extinct; and the rest, seed scales and branches dropped by araucariacean conifers. Many of the plants found fossilized in these two deposits are forms adapted to particularly warm, humid conditions. Moreover, silicified fossil coniferous wood found in early Cretaceous, shallow marine sandstones just north of George Grey's boat-wreck site in Gantheaume Bay possess pronounced growth rings. These are produced by trees that inhabit a distinctly seasonal environment – one perfect for silcrete formation.

For much of the Cretaceous Period, from about 160 to 75 million years ago, silcretes formed around the world with a vengeance. This is because it was one of the hottest and wettest times on Earth. The average global temperature reached about 24°c in the late Cretaceous. This is compared with about 14°c today. This was a time of active tectonic plate rifting and increased volcanic activity, leading to greatly enhanced atmospheric CO_2 levels. In Western Australia this was particularly so about 130 million years ago when India rifted off from the western margin of the Australian continent. Not only was this increase in CO_2 levels conducive to the production of chalk in the oceans, but it generated ideal conditions for silcrete formation on land. Cretaceous silcretes have been recorded on every continent, except Antarctica. While mainly centred on Cretaceous lower

latitudes, they are known from localities that were up to almost 60°
latitude in Cretaceous times, in both the Northern Hemisphere, in
Canada, and the Southern Hemisphere, the most southerly being the
site of Grey and Smith's rocky bed, just south of Gantheaume Bay.

High atmospheric CO_2 levels, high global temperatures and
intense chemical weathering characterized Cretaceous times. It has
been proposed that not only did this promote the formation of silcrete, but the prolonged high levels of silica in the terrestrial aqueous
environment was even a factor in the evolution of grasses.[21] These
plants generally take up silica in large amounts, as their normal
growth and development are dependent upon it. If silica is not
present, grass growth is stunted, and when grasses die they release
silica back into the soil, helping to maintain high silica levels in the
groundwater.

* * *

WHILE THE PLANT REMAINS entombed in the rocks tell of a time
when the world was a much warmer and wetter place, the silcretes
themselves can only form in such an environment that also has pronounced seasonality, one season very hot and wet, the other much
drier, but still warm. Rocks such as silcrete are a little like books,
containing the secret history of the climate at the time they were
being formed and, with the fossils they enclose, providing a snap-
shot into ancient environments. In the case of the history of the
sarsen silcretes of Stonehenge, there is a certain pleasing symmetry
in terms of their formation and subsequent use. These were rocks
produced by the power of the seasons being used, it would seem,
by the Neolithic inhabitants of England for their own particular
worship of the seasons.

Sumer Is Icumen In

Summertime, in a year shortly before the greatest mass extinction that life on Earth has experienced, some 252 million years ago. Known as the 'Permian–Triassic mass extinction event', it removed about 90 per cent of Earth's animal and plant species. But that summer, the saline lake that lies in what is now borderland between Texas and New Mexico is relatively tranquil, much like the thousands of summers that had preceded it and which were to follow. There is a set pattern to the type of sediments that are deposited in the lake each year – a pattern dictated by the cycles of the seasons.

Annual depositional layers of anhydrite (white) deposited 252 million years ago in winter, alternating with calcite (brown) deposited in summer, exposed near the Texas–New Mexico border on the side of u.s. Highway 62.

The clouds are building as strong summer low-pressure cells draw warm, monsoonal, moisture-laden maritime winds from the ocean Panthalassa.[22] This sea lies to the west of the supercontinent of Pangaea. It is here that the Castile Lake sits, adjacent to the coast and straddling the equator. Originally the lake had been an inlet, open to the sea, but this maritime link was cut off by the growth of a reef. Over time, the saline water steadily evaporated, resulting in the lake becoming hypersaline. It is large and deep: at about 25,000 square kilometres (nearly 10,000 sq. mi.) in area, the lake is about the size of Lake Erie, and its depth has been estimated to be around 500 metres (1,640 ft).[23]

As summer begins, the surface of the lake is 2 to 5 metres (6½–16 ft) lower than the adjacent sea. This follows a winter when strong high-pressure cells formed over the massive Pangean land mass, driving warm, exceedingly dry winds off the land to the sea, passing over the Castile Lake as they went. Extreme evaporation lowered the surface of the lake and calcium sulphate, in the form of the mineral anhydrite, crystallized and accumulated as a thin layer in its quiet depths. Summer moves on and marine water seeps into the lake through the extinct permeable reef, raising it back to its early winter level. It brings with it dissolved CO_2. This is eagerly exploited by algae growing in the warm water. With the extraction of CO_2, the pH of the lake decreases and instead of anhydrite, calcium carbonate, in the form of calcite, precipitates and settles to the lake floor. As it crystallizes, it mixes with the massive amount of algae that has been growing through the summer, turning the otherwise white calcite into a dark brown layer, covering the preceding winter's fall of white anhydrite. Despite the monsoonal nature of the weather, little freshwater is added to the system from inflowing rivers or from the summer rain. The hinterland is largely flat, so the lack of topographical relief inhibits appreciable rainfall.

A year: one white layer (winter, of course), one dark brown. This sequence of bands, known as varves, repeats over and over again – 190,000 times, it has been estimated. And 3.5 kilometres (2 mi.) northeast of the Texas–New Mexico border along u.s. Highway 62, the turning of the seasons can be seen played out in outcrops of dark brown and white striped rock. There is even some rhythm to the grouping of the stripes. Some are clustered in thin bands, barely 0.1 millimetre (¹/₂₅₀ in.) wide. Others are thicker, each up to 1 millimetre (¹/₂₅ in.). Each group forms a distinct 'millennial' cycle, up to 4 metres (13 ft) thick, representing periods of formation between 1,800 and 3,000 years in duration. The other distinctive feature of the banding is that, like a barcode, each set has its own characteristic pattern, meaning that groups of layers can be traced laterally for large distances, sometimes more than 100 kilometres (62 mi.).

It would be easy to think that such a rock type, forming so long ago and in such a distinctive way, was unique in the geological canon. But some of the oldest rocks on Earth, more than 3 billion years old, show that even then the turning of the seasons was just as potent an agent in influencing the nature of rocks. Known as the Buck Reef, a 250- to 400-metre-thick (820–1,312 ft) sequence of black-and-white-banded Archaean cherts outcropping near Barberton in Mpumalanga province in eastern South Africa provides compelling evidence for the influence of the cycles of the season on the nature of the rocks.[24] Formed about 3,416 million years ago in a shallow marine embayment, there was little sediment input from the land, suggesting a predominantly dry climate. Chemical sedimentation was the principal mode of accumulation.

Alternating dark and light cycles are present within the cherts, produced, in all probability, in a similar fashion to the hierarchical banding in the Archaean banded iron formation. In the case of the Buck Reef, it comprises alternating black and white layers, each roughly 5 centimetres (2 in.) in thickness. When examined in microscopic detail the black bands resolve themselves into 1,600 to 2,500 black-and-white couplets, each layer being 10–15μ thick. The couplets are thought to represent variable styles of deposition in summer and winter. During the warmer summer months there was active microbial growth providing organic material that combined with the crystallizing silica. Submarine volcanic activity was the most likely source of silica, injecting it into the marine system. During the cooler winter months, reduction in microbial activity meant that almost pure silica was deposited. On the larger scale, the cycling of the 5-centimetre-thick dark and light layers points to control by Milanković cycles of precession and obliquity driving glacial-like and interglacial-like conditions. In the cooler phases, silica was much less soluble in the colder water and precipitated more readily. The absence of fine dark micro-layers implies cessation of microbial activity even during the summer months.

Trees, Fungi and Calcretes

It is probably no exaggeration to say that calcrete is the most nondescript of all rocks. Imagine a lump of concrete: grey, extremely hard and, let's face it, excruciatingly boring to look at. Calcrete has much the same personality. Both are the products of organisms that inhabit this planet. The only real difference would seem to be that while the former is a construct of humans, calcrete, somewhat surprisingly, is manufactured by fungi. Without them calcrete just would not exist. But its story is not only about fungi. It is also about its close relationship with tree roots – not to mention its mysterious attraction to gold. However, fungi, roots and gold only come into play in the history of a piece of calcrete with the over-riding control of the passing of the seasons.

One of the few left standing – a rhizolith, about 5 cm (2 in.) long, formed of calcrete around a plant root, exposed by deflation of the sand in which it formed. Nambung National Park, Western Australia.

Compositionally, calcrete is the same as limestone: calcium carbonate. It is a rock that has many different names in different places – *caliche* in north America, *kunar* in India, *calcaire* in France, along with a glossary of other names in other places, such as *chebi-chebi*, *kafkalla*, *gitti*, *cornstone*, *harsua*, *beda*, *tepetate* and many more.[25] With calcrete (or whatever you want to call it) we step back once again into the realm of rock recycling. Many owe their existence to dissolution, generally by acidic groundwaters, of pre-existing lime sands and limestones. This enriches the groundwater with calcium bicarbonate and, given the right climatic conditions, it is recycled back into calcium carbonate in the form of solid, generally featureless, calcrete. Like silcrete this rock forms on land, often in soils. And, like silcrete, the role of the seasons in controlling the activity of groundwater is crucial to its formation.

Calcrete generally occurs as irregular masses in the soil horizon, though it can form extensive surface calcrete hardpans, such as in the Makgadikgadi Pan in Botswana and at Lancelin, Western Australia. It also forms in a more complex fashion in other subsurface environments, reflecting multiphase recycling of calcium carbonate generated by repeated cycles of climate change operating at different scales, from seasonal to millennial. Trees play a crucial role in its formation. It is a rock that is an evapotranspiration product of trees, rather than being just a specific adaptation of the trees to growing in the alkaline soils in which calcrete is present. In particular, it is intimately associated with the roots of trees and a whole host of plants.[26]

* * *

CLOSE TO THE COAST of Western Australia, barely 200 kilometres (124 mi.) north of Perth, thousands upon thousands of pinnacles of rock rise out of a sea of yellow sand. Individually they closely resemble the standing stones at Avebury and Stonehenge. However, these stand out from their Neolithic counterparts, both in their huge numbers and in being made of calcrete, rather than silcrete. Neither were they erected by humans; other organisms were responsible. Their construction was orchestrated by a formidable sequence of events, from annual to millennial climatic cycles, to the activity of plant root systems and their associated mycorrhizal fungi. The timing of the operation of all these factors has led to the production of a sequence of calcretes that have helped generate an astounding landscape.

Many of the pinnacles are human-sized, whereas some are giants, up to 5 metres (16 ft) tall. At the other extreme are localized fields of pinnacles, little more than knee-high. The shapes of the pinnacles are just as variable. Many are conical; others are cylindrical towers, some overtly phallic-shaped. The small pinnacles are often no more than simple stubs of calcrete – as though concrete had been poured into irregular holes, then hardened and exposed to the elements. But the large pinnacles are quite different. Within them they carry clues to the origins of these strange structures and a complex tale of changing climates controlling the fate of rocks.

What makes them stand out (both literally and figuratively) is that they appear to have suffered a rather bad concrete job: they are, for the most part, swaddled in calcrete. But unlike the small pinnacles, this is a surface veneer, the coating no more than a handspan thick. For many pinnacles, endless sand blasting has peeled off part of the calcrete to expose the limestone within. Often this comprises an entangled web of tubes, themselves made of calcrete – fists of arthritic stone fingers locked in a frozen embrace. These are rhizoliths (literally 'roots of stone'): the fossilized roots of plants that riddled the lime sands before they were cemented into limestone. The

Solitary pinnacle almost 3 metres (10 ft) tall showing an outer layer of calcrete wrapped around internal dune-bedded limestone, Nambung National Park, Western Australia.

intertwined fossil roots are sometimes overlain by remnant, gently inclined parallel layers – accumulations of beds of windblown sand, covering the ancient soils through which plant roots grew, eventually cementing to form the limestone that became the core of the large pinnacles.[27]

The story of the transformation of lime sand into pinnacles has its origins in Milanković cycles that generated the global glacial and interglacial cycles. For nearly half a million years off the western margin of the Australian continent, the Indian Ocean has risen and fallen roughly 130 metres (425 ft) with each 120,000-year-or-so glacial/interglacial cycle. During cold, glacial times, more water was locked in ice caps, lowering the sea level. As they melted during the transition into warmer, interglacial periods, the sea level rose. The ocean carried within it the broken remains of countless invertebrates that had lived on the seafloor: molluscs, starfish, sea urchins, bryozoans, corals, gastropods. Where the waves crashed on the shoreline, these carbonate-dominated sands were dumped unceremoniously, like refuse from a marine front-end loader. Wind built them into towering dunes, but once the sea began to retreat the stranded sand dunes were soon stabilized by plant communities. Their roots, while drawing water and nutrients from the sands, also played a crucial role in stabilizing the dunes and transforming them into limestone.

Up to 90 per cent of the sand grains were made of calcium carbonate; the rest were mainly quartz.[28] The Mediterranean nature of the climate produced pronounced seasonal differences in both temperature and rainfall, with hot, dry summers and cool, wet winters. This had a transformative effect on the formation of the calcrete. During the wet winter months, the dune carbonate sand suffered partial dissolution. It recrystallized as a carbonate cement as the dry summer months progressed, binding the dune sand grains together and transforming them into limestone made by the wind: aeolian limestone. Known as the Tamala Limestone, it snakes for about 1,200 kilometres (745 mi.) along the southwestern Australian coastline. Five Milanković cycles can be recognized within it.[29] Each consists of a thick layer, the aeolian limestone, overlain by a thin layer of calcrete. This sits at the base of a fossil soil horizon, formed during a period of stabilization of the dune when sea level had fallen as the world slunk into the next glacial cycle and the dune became stabilized by vegetation and turned into limestone. In the following interglacial period, sea levels rose as the ice caps melted. The shoreline once again migrated inland, bringing with it the next towering waves of sand dunes to engulf the stabilized vegetated surface. Climate cycle followed climate cycle. The result – cyclical sequences of the rocks. While the climate may have been the driving force, calcrete would not have formed without the mutual aid of tree roots and fungi.

A stand-out feature of much of the aeolian limestone is not just that the carbonate grains became cemented together by secondary carbonate cementation, but that recrystallization in a narrow zone around plant roots was intense and extensive. This was thanks to the symbiotic relationship between mycorrhizal fungi and plant roots. The plants provide the fungi with sugars formed by photosynthesis, the fungi provide the plants with nutrients, such as phosphorus, that they have sequestered from the soil. Many roots are confined to the upper couple of metres of the dune, though some will extend down more than 30 metres (98 ft) in search of water and nutrients. Fungal metabolism operating around the roots generates oxalic acid. Initially this facilitates dissolution of the carbonate, but precipitation of oxalate salts releases bicarbonate anions that scavenge calcium ions to produce secondary calcium carbonate. This concentrates as a zone around the roots, encasing them in a hard sheath of calcrete, which helps further

Tangled mass of rhizoliths, formed of calcrete around plant roots that developed in a calcareous sand dune, now cemented as an aeolian limestone, Cottesloe, Western Australia.

to bind the limestone.[30] An interweaving meshwork of rhizoliths is created, strengthening and binding the limestone together.

Calcrete has a somewhat unexpected relationship with gold, arising from this dependence on mycorrhizal fungi and plant roots. Research into methods of gold exploration has led to the realization not only that is there a close relationship between gold and calcrete, but that the association may form relatively quickly.[31] Anomalously high accumulations of gold (nearly one part per million) in calcrete associated with eucalyptus tree roots can form in the life span of a single tree. Gold concentrated in calcrete by tree roots can even form discrete nuggets more than 1 millimetre in size.[32]

Gold is incorporated into calcrete in two ways. Roots can absorb gold directly from weathered soil horizons, leaching into the calcrete that forms around the roots due to fungal and other microbial activity. Gold taken in by the roots can also pass up into the plant and be concentrated in leaves.[33] When the leaves fall and accumulate in the leaf litter, it decomposes, concentrating gold in the soil. This is then leached down into the lower levels of the soil profile where it can become absorbed in the calcrete that is forming around the roots.

* * *

THE RELENTLESS TURNING of these extreme seasons – cool and wet, hot and dry; cool and wet, hot and dry – ground on for thousands of years. But slowly, ever so slowly, the limestone and rhizoliths the seasons had themselves created were steadily and relentlessly destroyed. One of the fates of limestones of any age, in any part of the world, is to be subjected to the ravages of dissolution, slowly dissolving away to nothing. This often occurs in a very distinctive manner by the development of almost perfectly circular pipes that extend from the soil horizon down and through to the limestone below, slowly eating it away. They are called solution pipes. Generally extending to between 2 and 9 metres (6½–30 ft) below the surface, they can occasionally reach 20 metres (65 ft) in depth. In diameter they vary from about 20 centimetres (8 in.) to 1 metre (3 ft), and usually occur in dense groups – so close, in fact, that as they grow, they often merge with adjoining pipes. As the calcium carbonate

Pinnacles 2 to 3 metres (6½–10 ft) tall formed of aeolian Tamala Limestone, with outer protective layer of calcrete, Nambung National Park, Western Australia.

fraction of the limestone dissolves during their formation, the insoluble quartz residue accumulates in the pipes. Once they have formed, the rims of the pipes become encased in layers of calcrete, up to 10 centimetres in thickness.[34]

Such discrete pipes develop by focused dissolution of the somewhat porous aeolian limestone. How and why they form is due to a number of different factors. Foremost appears to be the role of trees. At a time when rainfall is high, stem flow of water down tree trunks or even the stems of shrubs will concentrate the water. This can cause extreme dissolution immediately beneath the plants, leading to the establishment and development of solution pipes. Once initiated, even after a particular tree has disappeared, solution pipes continue to develop. The frequent close proximity of solution pipes to each other points to extended periods of successive generations of trees contributing stem flow and pipe growth. Plant roots also play a part in pipe formation, providing channels through which water can percolate. Roots involved in facilitating the formation of solution pipes will sometimes also become encased in a straightjacket of calcrete, and can even cement the quartz infilling in the solution pipes. Surface depressions also determine where a solution pipe will begin to form. Being sites of accumulation of leaf litter, ponding water is likely to be more acidic and so more aggressive in dissolving the limestone.

In terms of long-term climatic changes, solution pipe development was favoured prior to the onset of glacial conditions, when rainfall was consistently high. Drier and more seasonal climates appeared during the glacial period, following solution pipe formation. This favoured the growth of calcrete rims lining the solution pipes. Smaller pipes were sometimes completely filled by calcrete. Like the calcrete of the rhizoliths, solution pipe calcrete growth was facilitated by the seasonal cycles of warm and dry, cool and wet. But its crystallization only occurred in the presence of the mycorrhizal fungi. As solution pipes increased in number and in breadth, many merged together, and all that was left were a few pillars of limestone, encircled by a protective coating of the calcrete that had once lined the solution pipes. Denudation of the vegetation in the last hundred years or so has exposed this complex world of limestone and calcrete buried in a sea of yellow quartz sand. The relentless winds – hot easterlies in the summer and cool, damp westerlies in the winter – have steadily blown this sand away, revealing a landscape of thousands of standing stones, a construct of the power of both millennial climate cycles and the turning of the seasons.

EIGHT

CENTRES OF ATTENTION

There is one group of rocks that consists of a number of rather strange bedfellows. One of this number owes its existence to rotting animal bodies; another to the labyrinthine meanderings of fungal mycelia; while the third, some say, formed under the influence of submarine earthquakes. Others blame bacteria. What they all share in common, however, apart from having rather strange modes of formation and their collective name – concretions – is their external appearance. They are smooth, rotund and, quite often, amazingly spherical. But above all, they are as hard as, well, stone. Very hard stone, that is. Mundane they may be on the outside, just dull balls of rock. Yet within some of these concretions are held secrets that in the past intrigued alchemists and Enlightenment figures alike. Some contain the evidence of wildfires that tore across the land millions of years ago, while others hold entombed the evolutionary history of some major groups of aquatic creatures. Similar they may be on the outside, but delve into these rocks and myriad stories are there, just trying to get out.

Centring in on Climate Change

Between the underworld of solid rock and the sky, there is a zone of rocks that are not always what they once were. This layer may be just the thinnest of veneers, but it may also reach more than 100 metres (330 ft) in thickness. It is a region of heavily altered rocks on the Earth's surface that have been changed due to the prolonged effects of the attrition of the atmosphere – especially its temperature and rainfall – chemically weathering the rocks over millions of years. A general term that is used to describe this region of weathered rocks, between

211

unaltered parent rock and the sky, is 'regolith'. Search any definition of this word and it will always be described as a zone of unconsolidated material – sand, silt, soil, gravel and the like. Certainly, this is often the case. But sometimes in this netherworld of the regolith, which rests between the underworld proper and the atmosphere, there are solid rocks that have been newly created by the action of Earth's atmosphere, recycling the very stuff of the underworld. These strange rocks contain within them some intriguing secrets.

Many rocks can be completely altered by the effects of chemical weathering. Even seemingly immortal rocks like granite and basalt can be transformed into something utterly different. For this transformation to take place, what is needed above all else is time. Huge amounts of it. Millions of years would help, aided by appropriate climatic conditions. As such, these rocks can provide an intriguing record of past climates and sometimes even show, within rock concretions no bigger than a quail's egg, evidence for past climate change.

The general name for the altered products of deeply weathered igneous rocks is laterite. This iron- and aluminium-rich material forms by intense weathering under a seasonally tropical climate, with periods of both high temperature and high rainfall interspersed with drier, though still hot, periods. The wetter phase speeds up the chemical decomposition of some of the mineralogical elements of the parent rocks, as well as enhancing the mobilization of the chemicals that are released into the groundwater. By dating the time when these altered rocks formed, it is possible to get a general idea of what the particular type of climate was in that region at their time of formation.

Deeply weathered laterite occurs widely on continents in tropical regions, between latitudes 35°N and 35°s. In parts of Africa, India, South America, Southeast Asia and Australia, it has been forming

Ferricrete pisoliths formed by the intense tropical weathering of granitic and gneissic rocks, Bedfordale, Western Australia. Pygmy sundew (*Drosera pulchella*) rosette about 15 mm (½ in.) across.

continuously for well over 100 million years on some of the ancient Precambrian terrains. It has even been proposed that parts of the very stable Yilgarn Block in Western Australia may have been exposed to such erosion for more than 500 million years. The thick weathered horizons that occur in some of these regions are dependent on a combination of the age of the landscape, levels of tectonic activity, climate history and the nature of the bedrock. Critically, the land mass must have been stable for extremely long periods of time. However, laterite also occurs in higher latitudes, for example in Oregon and Wisconsin in the USA and in Northern Ireland and Germany, where they are the product of slower, more prolonged periods of weathering under less intense tropical climatic conditions.

This lateritization process proceeds from the surface downwards, so the most intensely weathered material is nearer the land surface. Deep in the laterite, close to pristine rock, there is only minimal alteration of minerals by chemical weathering. Along joints and fractures in the host rock iron oxides are precipitated. This weakly weathered rock is known as *saprock*. As weathering progresses, most minerals (apart from quartz) break down and are transformed into a new suite of minerals, principally kaolinite (an aluminium silicate – otherwise known as china clay), gibbsite (aluminium hydroxide) and the iron oxides goethite and haematite, forming the more heavily weathered rock saprolite. With increased weathering all of the original textures of the parent rock are lost, and iron derived from mineral groups like pyroxenes, amphiboles and micas is redistributed through the weathering profile to form a hard, strongly cemented iron-rich layer near the land surface called duricrust (meaning, appropriately, 'hard crust'). The formation of the duricrust is driven by groundwater. Iron is leached from the weathered parent rock and, by subsequent evaporation, precipitates out as a hard crust of iron oxide minerals. As with silcrete, this is facilitated by seasonal wet and dry phases. Duricrust varies in colour from black to dark brown to red and is one of the more unusual types of recycled rock, both in terms of its formation and in the information that is held within it.

The lower part of the duricrust consists of angular, millimetre-sized fragments of almost-black haematite cemented together. But for the most part it consists of spheres or ovoids, ranging in size from about 5 to 20 millimetres (⅕–⅘ in.). These iron-rich spheres, of the size, appearance, but not consistency of a chocolate Malteser, are known as ferricrete pisoliths. They occur as either isolated spheres or cemented together. Colloquially they are known as 'pea gravel', and in many regions of the world they cover swathes of ancient Precambrian rocks. Unbroken, these pisoliths vary in colour from yellowish to reddish brown and are particularly hard. When broken open, they are mainly round or oval in section and show a variety of textures and colours, reflecting a variable mineralogy between pisoliths. The central core is often reddish brown to black. The black cores are formed of the minerals haematite, the magnetic form maghemite (of which more later) and gibbsite. The reddish brown cores lack the maghemite.[1] The pisoliths have an outer shell, or cortex, that varies in colour from pale brown to yellow and is mainly composed of gibbsite. Little haematite and goethite reach this zone.

Like plants and animals, fungi and bacteria, these pisoliths grow. To develop into such well-defined concretions, rather than being just a random mass of iron and aluminium minerals, they must form in an environment subjected to a suitable set of climatic conditions at the time of their formation. A number of different models have been proposed to explain why these minerals should grow as clusters of concretions. The most favoured is one based on observations of pisoliths formed in alternating humid and dry seasons in West African savannah climates.[2] Within the deeply weathered horizon of goethite and kaolinite, iron ions begin to migrate in response to variations in oxidation levels and differential drying due to variations in local porosity and seasonal variation in water availability. The iron begins to mineralize as tiny blobs of the hydrated oxide goethite. This process progresses by continued rehydration and replacement, a core of aluminium-rich goethite growing in a concentric, centripetal manner. While it is generally accepted that the role of seasonal hydration and drying out are crucial to the growth of the spheres, it has been discovered that there is another crucial player in this process: fungi.

When Robert Hooke peered expectantly down his microscope at the little calcite ooids from Ketton Quarry in the 1660s, he discovered a whole new world of mineral growth. Some 350 years later, geologist Ravinder Anand and his colleague Michael Verrall from CSIRO Perth, using an altogether more powerful scanning electron microscope, peered at the insides of little iron and aluminium oxide pisoliths from the Darling Range in southwest Australia.[3] What they discovered was not just a simple pattern of mineral growth, but evidence that various elements of the organic world had been at play.

Having formed in the upper part of the deeply weathered regolith, within metres of the land surface, the pisoliths show evidence of the activity of plants in their formation. Narrow channels wind through their mineralized centres. These are the pathways through which plant roots penetrated this horizon at the time of the formation of the pisoliths. And within the channels are very fine threads, barely more than a micron in diameter – the mycelia of mycorrhizal fungi that grew in symbiotic relationships with the plants. With the mycelia cluster spherical bodies, up to 10 microns in diameter, that are interpreted as fungal fruiting bodies.

The threads and fruiting bodies are now mineralized as gibbsite and various iron oxides and, based on radiometric dating of the pisoliths, provide evidence of fungi inhabiting this weathered zone between about 6 and 10 million years ago, that is, during late Miocene times.[4] Formation of pisoliths at this time is a direct indication that this region, which now experiences a Mediterranean-type climate, was subjected to a more humid, tropical climate. Rather than being just a passive presence within the pisoliths, Anand and Verrall have argued that the fungi have played a crucial role in the mineralization of gibbsite in particular, and in the formation of the entire pisoliths. Tiny crystals of gibbsite on and near the filaments show that the fungi

Block of cemented ferricrete pisoliths, *c.* 15 cm (6 in.) across, predominantly made of haematite, cemented by gibbsite. Formed by the intense tropical weathering of granitic and gneissic rocks, Roleystone, Western Australia.

directly influenced the formation of gibbsite in response to the highly aluminium-rich environment of these deeply weathered rocks.[5]

As well as being keys to the climatic conditions at their time of formation, other changes that some pisoliths in southwest Australia experienced tell of the drying out of the continent. In some, radiometric ages obtained from the centre of the spheres differ substantially from those taken from their outer crust, the inner core giving a much older age. Although it could be argued that these may represent just inordinately slow-growing pisoliths, the mineralogy of the inner core had changed, indicating intense heat from a single thermal event – a wildfire.

Maximum temperatures reached during wildfires and the duration of the heating event at the soil surface very much depend on the type of vegetative cover.[6] At a height of about 50 centimetres (20 in.) above the ground, temperatures can reach up to 1,100°C (2,012°F). At the soil surface they can be about 850°C (1,562°F). The effect of such intense heating on the mineralogy of pisoliths occurring very close to the ground surface can be profound. Many pisoliths that have a haematite centre also contain the minerals maghemite and corundum. This pair of minerals would not have formed during the initial growth of the concretion, but rather grew following the loss of the hydroxyl ion from original goethite and gibbsite when subjected to very high temperatures in the presence of organic matter. During wildfires, combustion of organic matter produces a partially reducing atmosphere, transforming haematite and goethite first to magnetite, then to maghemite as it cools. In addition to the changes in minerals during an intense wildfire, tiny apatite and zircon crystals present in the rock that are used to obtain radiometric (uranium–thorium/helium) ages can also be affected by any heating event in a different way. They experience a loss of helium, which affects the age of the surface of the pisoliths. Essentially the rock's clock is reset. All that is needed is an intense fire event persisting for 5 to 40 minutes in order for this to happen. Thus the age of the wildfire can be established.

While a lot more data are needed, one study of a pisolith containing maghemite with a 4-million-year-old core revealed that it had an outer crust about 2 million years old. This points to it having experienced a major wildfire at that time, and means that pisoliths may have the potential to be keys to understanding past climate change, unlocking details of the timing of the transition from the hot, humid climate that formed these tough little concretions to the much drier conditions during the Pleistocene – a time when wildfires began to have a devastating effect on the country. Each little pisolitic concretion can be seen, to all intents and purposes, as fossilized climate.

Yet there are other, much larger concretions that formed not on land but beneath the waves which owe their existence to a range of organisms other than fungi.

Cracking On

Nestled within John Woodward's five beautiful walnut veneer cabinets, housed today in the Sedgwick Museum in Cambridge, are many thousands of rocks, minerals and fossils. As was the tradition of the day, Woodward called them all 'fossils'. What we now recognize as the remains of once-living organisms he called 'Fossils that are extraneous'. Rocks and minerals he called 'Fossils that are real and natural'. Although Woodward collected much of the material himself, he also relied on an extensive network of contacts, both in England and overseas, to send him specimens. One, who sent him eight specimens, was Sir Isaac Newton. Renowned more for his mathematical prowess and his work on optics and physics, Newton also had some rather more arcane interests. One was alchemy. Although little was ever published in his lifetime, he wrote more about alchemy and the occult than anything else. The revelation that Newton was an active alchemist arose from research by John Maynard Keynes, an economist, following his purchase in 1936 of a large collection of Newton's unpublished manuscripts. Keynes's conclusion from what he found was that, rather than being the first of the age of reason, Newton was 'the last of the Babylonians and Sumerians, the last great mind which looked out on the visible and intellectual world with the same eyes as those who began to build our intellectual inheritance rather less than ten thousand years ago'.[7]

The three principal quests of alchemists in medieval and Renaissance Europe were to transmute base metals into noble metals,

Barely the size of a thumbnail, this specimen of cinnabar was once the property of Sir Issac Newton and most likely used in his alchemical experiments. He presented it to John Woodward and it remains in his collection in the Sedgwick Museum, Cambridge.

especially gold; the creation of an elixir of immortality; and panaceas to cure illnesses and diseases. To achieve any or all of these aims, the basic substances used were a variety of minerals and rocks. The eight specimens that Woodward acquired from Newton were all fundamentally important components of any respectable alchemist's toolkit, which was presumably why Newton had them in his possession in the first place.

Three pieces of cinnabar, two collected from Augsburg in Germany, one of which 'has but little Quick-Silver [mercury] in it: but may hold about 1/16 Silver, with a little Gold'; the other 'Out of a River in *Hungary*', were passed on to Woodward from Newton.[8] For Newton to have had this mineral was not surprising, given his deep interest in alchemy. Cinnabar is a sulphide of mercury, and these two elements played a fundamental role in alchemical experiments, both to transmute liquid mercury into gold, and also in the quest for the mythical 'Philosopher's Stone'. The medieval philosopher Roger Bacon (*c.* 1219/20–*c.* 1292) contended that 'red elixir', or the

Philosopher's Stone, was made of sulphur and mercury, the constituent elements of the deep red cinnabar. The principal use of this mineral in alchemy was to be heated and converted into liquid mercury. Processing cinnabar to produce mercury was a highly dangerous procedure. When burnt, noxious sulphur was driven off, leaving pure, toxic mercury. It has been argued that Isaac Newton indulged in this activity to such an extent that it brought on many of the ailments from which he suffered, and which are typical of mercury poisoning, such as depression, insomnia, mood swings, outbursts of temper and poor digestion.[9]

Newton also gifted Woodward a specimen of the iron sulphide pyrite, one, the 'common tesselated Pyrites, now of a Rust-colour. From . . . *Saxony*; whence twas sent to Sir *Isaac Newton*, as the *Ludus Paracelsi*.' Philippus Aureolus Paracelsus (born Theophrastus Bombastus von Hohenheim, 1493–1541), the Swiss alchemist, physician and chemist, was one of the first to promote the use of chemicals and minerals in medicines. Although an alchemist, Paracelsus was less concerned with the more occult side of alchemy. 'Many have said of Alchemy, that it is for the making of gold and silver. For me such is not the aim, but to consider only what virtue and power may lie in medicines.'[10] He coined the term *ludus* for pyrite, in allusion to its frequent occurrence as perfect cubes, like dice. Paracelsus believed that it was efficacious when suitably processed in treating ailments such as bladder stones (urinary calculi) – as he put it, 'stone against the stone.' 'Paracelsus', Woodward wrote, 'represents this Body as capable of dissolving the Stone in the Kidneys and Bladder.'[11] The quest for such a cure was driven by the incredibly high incidence of this condition at the time. It has been estimated that between the sixteenth century and the first half of the eighteenth century in northwestern Europe, it was responsible for perhaps one-third of all deaths.[12]

The main proponent of the use of *Ludus paracelsi* to treat internal stones was the Dutch chemist Jan Baptist van Helmont (1588–1644). Specimens sent in 1675 to the secretary of the Royal Society, Henry Oldenburg (*c.* 1619–1677), from an Antwerp apothecary, A. Boutens, purported to be the *ludus* used by van Helmont. This piqued the interest of the society, especially Robert Boyle, as well as Newton. While it was generally accepted that van Helmont's *ludus* was pyrite, confusion arose over further material that was brought to England by van Helmont's son, Francis Mercurius, who asserted it was used by his father for treating stones. Some were clearly not pyrite: no metallic sheen; no dice-like shape. Here were stones like very hardened clay, some with yellow-brown crystals lurking within. Specimens of this material he gave to Newton, who, many years later in 1720, after his interest in alchemy had waned, passed them on to Woodward.[13]

John Woodward's views on this 'Ludus helmontii' were transformational. Being one of the first people with an interest in geology to study rocks in an analytical, deductive manner, Woodward saw Newton's *Ludus helmontii* not as a potential miracle cure for bladder and kidney stones, but as little more than a fragment from a large type of rock concretion with which he was very familiar, having collected some himself from the foreshore on the Isle of Sheppey. Such rocks were, at the time in England, called 'waxen vein'. Now they are known as septarian nodules or concretions, or simply septaria. Forming anything from spherical, or semi-spherical, to more flattened 'Turkish-bread'-shaped concretions, they are composed of the same clay in which they occur and in which they grew. Typically the clay is cemented into concretions by the precipitation of calcite. What really characterizes septaria is their internal structure. They are fissured by cracks radiating from the centre towards the outer margin, often failing to reach it, and so externally these rocks look like little more than a dull, grey boulder. Sometimes they do, however, giving the surface a crusty, reticulate pattern. The fissures themselves are generally filled by coarsely crystalline calcite, varying in colour from brown to yellow.

How septarian concretions formed has been the subject of conjecture since Woodward first mused upon the subject. It was a rock that fascinated him perhaps more than any other. In his catalogue he set out his logical, scientific appraisal of what the rock was and how it could have formed. He did this when the underlying explanation for the formation of rocks in general was that they were a product of the universal biblical Flood. Like all other writers at the time, Woodward subscribed to this view, but its influence on his really quite modern scientific analysis of how such concretions formed was not overreaching.

First, he listed a number of 'reflections', which were his observations. These included 'the Bigness of this Body'; whether it was 'in Form of a Nodule, loose and independent'; that in the cliffs he found them 'lying flat-ways, and parallel to the Site of the Strata'; 'the Cracks in this Body' and 'the Constitution of the Partitions'. He observed how 'their growing gradually less, as they approach the Crust' as well as the infilling 'Crystallizations, & Efflorescencys.' Significantly, he observed the 'Partitions passing the Bodies of the Sea-shells dividing and parting them' and that 'those partitions . . . divide and intercept others.'[14]

Making allowance for the influence of the universal Deluge in his thinking, some of his deductions from his 'reflections', provided in true scientific fashion, were not too far off the mark. However, rather than realizing that the concretions formed within the clay after they had been deposited, he believed that 'The main . . . Mass of this Body concreted, and was form'd in Water.' Further, 'Upon the Retreat of the Water [of the universal Deluge] it settled down along with the Clay

... that form'd the Strata in which it was log'd.' He then correctly surmises that the 'Cracks in it were form'd afterwards.' Then the 'Cracks ... were, generally, in tract of Time, gradually filled by Spar [calcite]; the Water which is continually pervading the Strata deriving thence loose

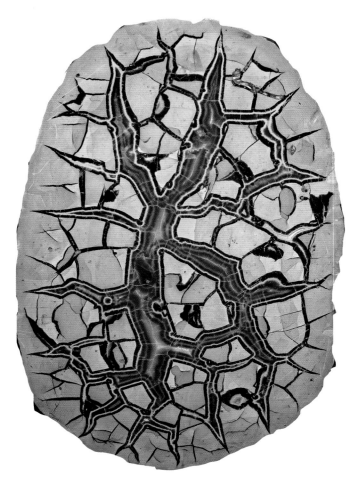

A cut-and-polished section of a septarian concretion, showing the characteristic fractures infilled by calcite. In the Pinch Collection at the Canadian Museum of Nature, Ottawa, Canada.

Particles of that Mineral, introducing them into the Cracks, and affixing of them there, so as thereby to form the Plates and Partitions.'[15]

As a first pass in explaining a rather curious geological phenomenon, Woodward's explanation is not unreasonable. In more recent times, attention has centred on what factors drive the concretions to form and why they crack. Woodward was on the right track when he wrote that 'wherewith the Pores and Interstices of the Body were saturated, during the Coalition of the Matter that compos'd it, gradually quitted it, and got forth, the said Matter was contracted, and shrunk upon divers Directions . . . and the Mass parted into Segments or Tali'. In other words, as the concretion de-watered, it shrank and cracked, allowing calcite to fill in the spaces.

The concretions with which Woodward was familiar, which are now known to have come from the Eocene (about 50 million years old) London Clay, are not overly large, generally up to about half a metre (20 in.) in diameter, and rarely truly spherical. But on the other side of the world, in slightly older clays, are found the most spectacular of septarian concretions.

* * *

IT IS 1848, on the east coast of the South Island of New Zealand. Walter Mantell has been working his way down the coast, making notes on many of the geological phenomena he is seeing. Given his heritage, this is no surprise. As the son of Mary and Gideon Mantell, discoverers at Tilgate Forest in Sussex of some of the earliest evidence for dinosaurs, his upbringing had been swamped in the rocks and fossils that packed his parents' house in Lewes. Eight years earlier, he had left England to travel to the other side of the world. But his reason for travelling down the coastline of South Island was not primarily geological.

Mantell had been appointed as 'commissioner for extinguishing native titles' on South Island and had been instructed by none other than George Grey, by then governor of New Zealand, to negotiate the establishment of native reserves with the main Maori tribe of the South Island, the Ngai Tahu. He offered them the miserly amount of 4 hectares (10 ac) per head of population, and this derisory amount had been resented by the Maori tribe, especially as Mantell had also, on Grey's behalf, promised them that the government would provide schools, hospitals and general care. None of this ever came to pass and the guilt he felt from this failure to treat the Maori tribe fairly is said to have haunted Mantell for the rest of his life.

Of the various geological discoveries made in the region by Mantell, arguably the most spectacular were on a beach at Moeraki, on the east Otago coastline. In the detailed notes that he sent back to

The largest of the septarian concretions, some more than 2 m (6½ ft) in diameter, washed out of the cliff at Moeraki, New Zealand.

his father in England, who read them before the Geological Society of London in 1850, Mantell records how in the cliffs:

> Midway between the Bluff and Moeraki, the clay contains layers of septaria, varying from one to five feet and more in diameter. Hundreds of these nodules, which had been washed out of the undermined clay cliffs by the encroachment of the sea, were scattered along the beach . . .[16]

Many of the large concretions are still there today. Often virtually spherical and exceeding 2 metres (6½ ft) in diameter, it has been estimated that some weigh up to 7 tonnes. One study has estimated that they grew extremely slowly, taking as long as 4 million years to attain that size.[17] The growth of the concretions took place within the 60-million-year-old clay that forms the cliffs.

Septarian concretions are found worldwide, and most share the same form, texture and mineralogy – clays cemented by calcite and containing radiating cracks and fissures partially or completely infilled by calcite of various colours. Although previous suggestions for how they formed largely follow Woodward's model of dewatering, causing shrinkage and cracking, one alternative proposal has raised the possibility of submarine earthquakes being responsible for the cracking.[18] But it has taken a detailed study of septarian concretions from Jurassic rocks exposed on the northeast coast of the Isle of Skye in Scotland to reveal that, once again, the formation of these rocks was controlled almost entirely by the activity of organisms; in this case, bacteria. Specifically, by bacterial organic-matter degradation in muds prior to their lithification.[19]

Outcropping on the foreshore are dark grey claystones of the Flodigarry Shales, deposited about 155 million years ago in a marine setting subjected to periodic stagnation events and oxygen depletion of the bottom waters.[20] Set within these claystones are very hard, subspherical concretions, up to about half a metre in diameter.[21] When sectioned, all are suffused by radiating cracks, filled by white, yellow and/or brown calcite. The research suggests that hardening of the clay into discrete concretions, along with internal shrinkage, opening of cracks and the beginning of the growth of the septarian calcite within them took place virtually simultaneously and were inter-related. Growth of the concretions took place at shallow depth in the clay, during periods when deposition had ceased. Rather than having a prolonged period of formation, as has been suggested for the Moeraki concretions, evidence from isotope analysis suggests rapid formation of the hard concretions. This, it is argued, was caused by the rapid precipitation of the calcite cement and was probably associated with bacterial colonization of the sediment.[22] This involved bacterial sulphate reduction, transitioning to methanogenesis.

Development of the septarian cracks is thought to have been intimately related to the properties of bacterial colonies. Shapes of the cracks reflects tensional rupturing during the hardening of concretions. Up to 43 per cent of a concretion can be occupied by cracks, suggesting that the concretions at the time of crack formation were only weakly hardened. The conditions that best produced these structures are thought to have been caused by the biodegradation of wet, gel-like flocculated clay containing extracellular polymeric substances (EPS). This sticky mucus formed from bacterial excretions and dead bacteria was derived from bacterial colonies inhabiting the upper part of the sediment. Further bacterial decay transformed these EPS gels into polysaccharide fibrils that bound the surrounding clay.[23] Significantly, this process released calcium ions that were involved in the calcite cementation. As the EPS degraded, organic molecules were released and taken up by the crack-filling calcite. The development of a dark-brown calcite colour is one such effect. Organic material containing fatty acids, on the other hand, caused a yellow-coloured calcite to form, subsequent to the brown one.

Given the similarity in form and mineralogy of septarian concretions around the world, it is not unreasonable to assume that similar processes were involved in their formation elsewhere. One intriguing, but essentially unanswered, question is why did the concretions choose to form where they did? Diffusion of calcium to form the calcite that cemented the clay in concretions would have come from a single point. The most probable cause was greater activity of bacteria, causing the decay of some particular piece of organic matter. There are other types of concretions, however, that can form in clays that lack shrinkage cracks and the glittering coruscation of calcite crystals. Rather, at their centre, as though lying in a self-made tomb, are the complete remains of animals – bones, blood vessels, nerves. Everything.

Close to the Bone

It is possible to stand atop the core of the 370-million-year-old Great Devonian Barrier Reef in the southern Kimberley region of Western Australia, and then to take a metaphorical plunge into the deep fore-reef sediments. No scuba gear necessary here, just a pair of stout walking boots. As you descend into what was once deep water, but is now a rather steep, rocky slope, it is possible to follow a single bed of limestone and see how it subtly changes – in colour, in texture and in the fossilized remains of creatures that once swam here, living on and within the sediment that cascaded off the reef. Reaching further and further from the reef core, into what once was water hundreds of metres deep, the slope has gone, and you walk across a wide plain. Now it is lined with little cobbles. In size and shape, think squashed bowling balls; in colour, pale grey, but weathered from very dark, almost black, shales. You have reached the Gogo Formation. This is not a place you would wish to have been in Devonian times. No light and no oxygen. Down into this cold, black mud dead bodies sank from above – fishes and crustaceans, primarily, that once swam in the shallow waters of the reef. As soon as their carcases were swallowed by the sediment, they became prey for the only life that could inhabit this dark world: bacteria.

Many of the fishes that swam around this Devonian reef would not have looked too out of place in modern seas: ancestral sharks, although many only the size of robust goldfish; pilchard-sized teleost (bony) fishes; lungfish, rare today, but not then. Like most modern fishes, teleosts and lungfishes had a full complement of bones encased in soft tissue and a scaly skin. But the most common of the Devonian fishes were altogether different. These were the placoderms – fishes that wore their bone on the outside.[24] Covering their head and the anterior part of their body trunk, this armoured bone was an anatomical feature that, indirectly, was responsible for the astounding preservation of these fishes within the hard limestone tombs that shrouded their bodies.

Extracting fossil fishes from these hard limestone concretions that accumulated in the spinifex-covered outwash plains away from the main limestone reef front can be geologically challenging. The crude way in the field is to place one's foot on a likely looking concretion and, should it be thought to contain a fish, hit it very hard with a hammer. More often than not, nothing appears as the rock splits in two. Yet now and again a whole fish is exposed to the light of day for the first time in hundreds of millions of years. But it is possible to extract an

A limestone concretion from the Late Devonian Gogo Formation, Kimberley region, Western Australia, formed around the body of a decaying lungfish, *Griphognathus*. The skeletal elements are all original bone.

entire three-dimensional fish from these concretions with a rather less brutal technique. This more sensitive, though time-consuming, way is to soak the concretion in weak acetic acid, like vinegar. It may take many weeks, but because the limestone concretion is made of calcium carbonate, it slowly dissolves away to reveal the bones of the fish. These bones survive this chemical ordeal because the concretion formed so rapidly around the fish that the calcium phosphate of its bones was preserved and is impervious to the effects of the acid. The same method has been used on another astonishing fossil fauna of Cretaceous fishes that are preserved in similar concretions in the Santana Formation in Brazil. Like the Gogo fishes, the shape of the Santana fishes mimics the shape of the entombed fish.

For many years during acid preparation of the Gogo concretions, undissolved residue was simply washed away down the sink. This was until it was realized that not only had bone been preserved in the rock, but so too had an astounding array of soft tissue, much of which had simply been lost during preparation. Once this had been understood, more careful acid preparation of concretions yielded specimens of fish that not only had their original bone preserved, but muscle, circulatory and nerve tissues. The reason why such soft tissue could be fossilized was that the fish would have been subjected to extremely rapid phosphatization of this tissue. It probably took place within a few days, or even hours, of the fish sinking into the mud. The discovery of the soft tissue was made in some of the placoderm fishes, the preserved soft tissue being tucked up against the bony armour plates of the body. Such a closed environment created a suitably protected location that was conducive to rapid replacement of the soft tissue by calcium phosphate.[25] Here was an anoxic environment where CO_2 accumulated and the presence of volatile fatty acids reduced the pH, so initially favouring precipitation of calcium phosphate, rather than calcium carbonate. The presence of microbial film on the tissue may have also concentrated the phosphate, which had leaked out of the fishes' bones.

The variety of soft tissue found preserved in these concretions is astonishing. In addition to muscle, blood vessel and nerves, one small specimen of a placoderm, barely as long as my little finger, was found to contain tiny bony plates nestled in its body cavity. Initially they were thought to have been the remains of the fish's last meal. However, detailed study revealed them to be plates of a single intra-uterine embryonic fish.[26] Moreover, extending to the area where the tiny plates are concentrated is a thin tube. It is believed to have been the mother fish's umbilical cord. Perhaps even more remarkable is the presence of an amorphous crystalline mass that may possibly represent the embryo's yolk sac. These fish were indulging in internal fertilization and giving birth to live young.

Like many other similar concretions that contain fossils, those of the Gogo Formation only formed because of the bacterial decay of the bodies of the animals that took place in the soft, black mud. In their death, with the aid of bacteria, the decaying bodies of the organisms made their own sarcophagi. Without life, these rocks would not exist. This is because the carbon that became fixed in the calcium carbonate concretions came mainly from the organs of the organisms inside the concretions as they decayed after death. To have preserved soft tissue in such exquisite detail, the formation of these concretions must have taken place very soon after the animal's death.

As they began to decompose, carbon was released that became incorporated into bicarbonate ions. As their concentration began to rise sufficiently, a reaction front moved outward, providing conditions suitable for the precipitation of calcium carbonate that cemented the mud into a concretion.[27] When the reaction front was just developing from a very small amount of organic material, concretions formed that were spherical. However, as the elongate body of a fish decayed, it formed a concretion that mimicked its body shape. Micro-pores within the forming concretion functioned as migration paths for the bicarbonate ions. These combined with calcium ions in the pore water to form the calcite that bound the sediment around the fossil and formed the concretion. Growth continued until all the organic carbon produced by bacterial decay was consumed. The concretion formation then stopped.

* * *

THE IMPORTANCE OF ROCK concretions to our understanding of the evolution of life and to the nature of past ecosystems should not be underestimated. The propensity of organisms under the right conditions to create their own sealed rock tombs provides an intimate glimpse into ancient worlds. Arguably the most important deposit of concretions containing exceptionally preserved fossilized animals and plants of high diversity and abundance is the so-called Mazon Creek Lagerstätte.[28] The exceptionally preserved fossils have been collected from the Mazon River area in Illinois for about 150 years.[29] Collecting intensified in the 1940s with the advent of extensive strip-mining for coal. These concretion-bearing sediments that overlie the coal seams are late Carboniferous in age, close to 310 million years old. The concretions are found in the lower 3–8 metres (10–26 ft) of the Francis Creek Shale that overlies the coal beds. Here the sediment consists of rhythmically alternating cycles of silt and clay, thought to represent deposition in neap–spring cycles. The concretions that occur within them are no limestone concretions. Rather, they are made of the iron carbonate mineral siderite.

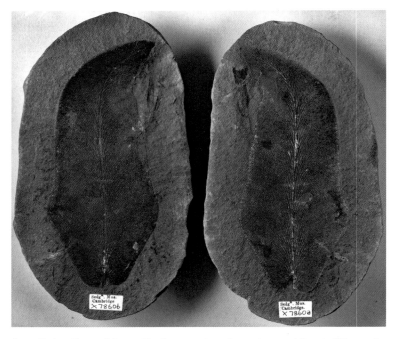

Late Carboniferous Mazon Creek concretions formed around frond of *Pecopteris*, a tree fern, 310 million years old.

An astonishing number of species has been found in the concretions in these shales. Like a geological museum formed 310 million years ago, this layered collection of concretions holds within it an intimate insight into a late Carboniferous world – of animals and plants from the land, and animals from the river and sea. Washed down the river into the sea were plants, millipedes, centipedes, insects, amphibians and a reptile. Many soft-bodied marine animals also became encased in concretions – jellyfish, sea anemones and worms of many kinds. A panoply of marine invertebrates is represented, including crabs, lobsters, shrimps, barnacles, chitons, gastropods, bivalves, cephalopods and holothuroids (sea cucumbers). In all, about 350 species of plants and more than 465 species of animals, including some forty species of fish, have been described from the concretions.[30] This high diversity of organisms is reflective of sourcing from a wide range of habitats. The silts and clays from which the concretions formed were deposited in a deltaic environment into which representatives of plants and animals were sourced from coastal terrestrial swamps, as well as near-shore and fully marine environments.

In the deltaic setting in which the Mazon Creek sediments were deposited, there was a reasonably constant supply of sediment, predominantly from upland, peaty forests. This generated an environment in which organic matter and dead bodies were rapidly buried, effectively reducing their chances of being scavenged. It also meant that there was limited time for aerobic decay.[31] The cyclical deposits of silts and clays provide evidence for strong tidal flows. This would have limited the abundance of burrowing organisms that were unable to survive in such a mobile environment, producing low-oxygen conditions within the sediments and favouring the precipitation of iron carbonate, rather than calcium carbonate. Water sourced from upland forests decreased acidity, further enhancing the precipitation of iron as siderite.[32]

Due to the mud in which the concretions formed being essentially anaerobic, soft-tissue decay was retarded. In this oxygen-depleted system, anaerobic bacteria thrived. Their activity led to methane fermentation, which had the effect of initially cementing the concretion by facilitating the formation of the iron carbonate by infilling pore space in the sediment. As a consequence, it protected the organic remains that were responsible for the generation of the carbonate, preserving hard parts and, frequently, soft tissue as the concretions rapidly grew. Even though many of these rocks lack obvious fossil remains within them, the sediment was in essence saturated by organic remains. Their decay created environmental conditions in the sediment that triggered the precipitation of the siderite, even without an actual dead body.

While the geochemical setting of the sedimentary system obviously played an important part, the main constructors of the Mazon Creek concretions, like those of the Gogo Formation, were the animals and plants decaying in the sediment. They were greatly helped by suites of anaerobic bacteria that thrived in these sedimentary conditions. Without organisms these rocks just would not exist.

NINE

FOSSILS, FORESTS AND FIRE

They are both unprepossessing cliffs. True, one, in eastern Australia, has a little height to it. Sixteen metres, to be precise. The other is barely half this size and is partly hidden by the forest of the Ardennes in Belgium. Images of cliffs usually conjure up thoughts of vertical walls of rocks, plunging vertiginously into a raging sea beneath. These, however, are, to be kind, rather muted in their grandeur. They are the kind of cliff that, with a bit of energy, you can scramble up and have a good chance of reaching the top unscathed. As rock outcrops go, both are, for the most part, rather grey, reflecting the nature of the rocks. So, seemingly uninteresting, rather dull cliffs. Yet within their drab demeanours are hidden in the chronicles that tell of the extinction of great forests. Of cataclysmic wildfires, changing the course of evolution; of vibrant ecosystems laid to waste in palaeodystopian worlds. But they also reveal eventual recoveries to some semblance of normality. Such is the power of stratigraphy – documenting sequences of rock strata and unravelling their ancient narratives. Never write off rocks, the ultimate keepers of Earth's deep history.

When Louis XIV stopped to admire those mantelpieces in his palace at Versailles that had been so beautifully crafted from Rouge de Languedoc 'marble', he would have been quite unaware of the singular significance of this red limestone to one of the major events in the evolution of life on Earth – the disappearance of up to 75 per cent of marine animal species in the Late Devonian Period about 372 million years ago, during one of Earth's five great mass extinctions. But this period in Earth history was not all about loss, for it was also a time on land when forests had undergone their first dramatic expansion. The impact of this on the evolution of life, and of Earth itself, was to be profound.

Comings and Goings

Along the northern margin of the great southern supercontinent of Gondwana 385 to 360 million years ago, warm, shallow seas accumulated blood-red carbonate muds and silts, sometimes in association with fringing reef systems. Far to the east of the Montagne Noire quarries in Languedoc that provided Louis XIV with his ornamental stone, this same northern Gondwanan marginal marine environment sported a series of fringing reefs, preserved as a sequence of limestones in what is now the southern Kimberley region of Western Australia. It was the Great Barrier Reef of its age. Many of the rocks are fine-grained, mainly red carbonate siltstones that were once sediment shed into deep water on the seaward side of this great reef that fringed the Precambrian land mass. From Languedoc to Western Australia the calcareous mudstones and siltstones contain a rich fossil fauna of marine invertebrates.

Fossils provide a rich source of information when unravelling the history of rocks. As well as encapsulating many of the environmental conditions during which they lived and during which the sediments in which they are preserved were formed, patterns and processes of evolution and extinction can also be revealed. Among the many different types of fossils found in the Late Devonian sedimentary rocks of this north Gondwana province are thousands of exquisitely preserved trilobites.[1] Separated then, as now, by thousands of kilometres, the French and Australian rocks contain many trilobite species in common. As trilobites go, they are small, some as large as an adult's fingernail but others barely the size of the nail of a newborn baby. Yet what they lack in size they more than make up for in terms of the insights they provide into the biodiversity, evolution and extinction history of animals that once scuttled around in these ancient seas.

Now extinct, trilobites were arthropods – animals with a mineralized outer skeleton and related to insects, crustaceans and spiders – that grew by periodically moulting. Each mineralized moult had the potential to be fossilized, so a single individual could potentially provide dozens of fossils. All trilobites possessed a head shield, segmented body and platy tail. Resembling modern day slaters or woodlice, they were very diverse and an important part of the marine fauna in the period leading up to the Late Devonian mass extinction. The northern Gondwanan carbonate rocks were deposited as sediments both before, during and after the mass extinction event. As such, they record the extent and nature of the trilobite extinctions. The character of the rocks may also hold a major clue as to what caused this biologically catastrophic event, for the evolutionary changes occurring within the trilobites appear to have been the outcome of changes that were occurring in the ocean as a direct result of events on land – in particular in the newly burgeoning forests that were flourishing close to the coast.

Five major groups of trilobites lived in the warm, shallow Gondwanan seas before the extinction event. Some, called odontopleurids, were very spiny; others, the harpetids, had a head surrounded by a bizarre snowshoe-like fringe; a third group, the phacopids, possessed compound eyes with very large lenses; and a fourth, the scutelluids, were typified by having a tail that was as big as its head. The fifth group, the proetids, had a simpler, generalized trilobite form. Trilobites evolved about 540 million years ago, and

The ridge in the distant foreground is McWhae Ridge, in the Kimberley region of Western Australia. Close to the top the rocks record the Late Devonian mass extinction event. Here three major orders of trilobites became extinct, as they did at this level worldwide. To the left of the ridge is the main reef front off which the sediments were shed into deep water.

all became extinct 252 million years ago at the end of the Permian Period. This was the greatest of the five mass extinctions that have punctuated the evolution of life on Earth during the last half a billion years. But the Late Devonian rocks in Languedoc and the Kimberley show that this was also a bad time for trilobites. Of the five major groups that had existed since trilobites first evolved in the Cambrian Period, some 150 million years earlier, three became extinct at the Late Devonian mass extinction event. Of the remaining pair, one, the phacopids, lurched on for about another 10 million years until it too gave up the evolutionary ghost. Only the unspecialized and morphologically conservative proetids survived until the end of the Permian Period.

Some rocks can tell the time. This is time measured not by any metronomic considerations arising from radiometric dating, but, as recognized by William Smith early in the nineteenth century, by the comings and goings of fossil species recorded in the rocks. This reflects the evolution of organisms whose fossilized remains are now encapsulated within the rocks. For example, a certain type of fossil trilobite is found in the quarry at Coumiac in Languedoc. The same species is found in rocks thousands of kilometres away in Western Australia. The rocks in which they occur therefore formed from sediments that were deposited at the same time.

In the Late Devonian seas along the northern margin of Gondwana, changing time is seen particularly in the compound eyes of the trilobites. Not only did their bodies become smaller over time, but so too did their eyes and the numbers of lenses they contained. Some even became blind.[2] Evolution at work. Trilobites from each of the five major groups were affected, showing that there was strong evolutionary pressure from the environment driving these changes in all trilobites. The question this raises is whether these changes were a major cause of the trilobites' extinction, or just an expression of their response to changing environmental conditions that ultimately led to the mass extinction. The answer, it transpires, can be found in two places: first, in the colour of the rocks, and second, and perhaps somewhat surprisingly, in some butterflies placed on a starvation diet.

Many of the northern Gondwanan mudstones and siltstones from Languedoc to the Kimberley are blood-red in colour. This is due to the presence of the iron oxide mineral haematite. Microscopic examination of thin sections of the rock reveals that the haematite is

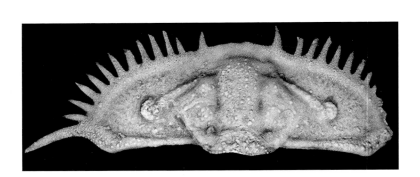

Trilobites collected from McWhae Ridge in the Kimberley region of Western Australia. These trilobites are members of two of the three orders that became extinct during the Late Devonian mass extinction event. Top is the headshield of the odontopleurid *Gondwanaspis dracula*. Bottom, headshield of the harpetid *Eskoharpes palanasus*. The block also contains tiny, bean-shaped ostracod crustaceans and the bivalve *Buchiola*.

not randomly spread throughout the rock, but rather is particularly concentrated in very thin bands – the remains of microbial mats that once carpeted the seafloor. Some fossil fragments also have bush-like growths of haematite, mini-microbialites formed from iron bacteria causing precipitation of iron oxide from dissolved iron in the seawater. This is the key to the environmental conditions in which the trilobites lived, for such bacteria flourish in environments very low in nutrients.[3]

Experiments on a suite of modern arthropods have revealed that reduced nutrient availability results not only in reduced body size in the offspring, but a concomitant relative decrease in size of 'nutrient-hungry' organs. Principal among these in arthropods are the eyes. In living arthropods, the development of large eyes is metabolically costly, in terms of both their growth and their operation. For instance, in some blowflies, just keeping photoreceptor cells ticking over while resting takes up about 10 per cent of the resting metabolic rate.[4] Evolution, then, becomes a fine balance between performance and the low cost of visual production. Experiments on individuals of the North American orange sulphur butterfly, *Collas eurytheme*, reared in a stressed, nutrient-poor environment show that their maximum body size was reduced.[5] Not only that, but they had correspondingly smaller compound eyes. And so with the trilobites. Evolutionary trends in reduced body and eye size reflect adaptations to progressively increased impoverishment of nutrients.

The mass extinction event during the Late Devonian that saw the extinction of so many species is shown not only by the major decline in species, but by changes in the character of the rocks. These preserve telltale environmental changes that were taking place as the sediments were being deposited and which were likely to have contributed to the extinctions. In the case of the Late Devonian mass extinction, both the drastic changes in the fossil faunas and, in some

Tiny microbialites from the Late Devonian of McWhae Ridge, Western Australia, made by iron bacteria growing from thin microbial mats and precipitating the iron oxide haematite. These microbes thrive in low-nutrient, low-oxygen environments. This thin section is about 1 square cm.

places, abrupt changes in rock types reveal that the mass extinction was not a single event. About 372 million years ago, and separated by about 400,000 years, occurred what are known as the two Kellwasser events. The second of these was by far the most intense. These events mark short periods when the sea level dropped and there was widespread deposition of black, organic-rich muds, now preserved as mudstones and shales in many parts of the world. Such muds reflect deposition during periods of eutrophication – a time when a spike in nutrient input into the sea led to huge algal blooms, resulting in oxygen being sucked out of the marine system. In many places, carbon accumulated in the anaerobic muds, turning them black. For an ecosystem adapted to low-nutrient conditions, these two events were, for many species, terminal, causing the extinction of many organisms adapted to a low-nutrient environment. But from where did the nutrients come? To trace their source, it is necessary to examine what was happening on the land at the time, and in particular the role played by the burgeoning first forests.[6]

Forest Fires

The Late Devonian mass extinction, which saw the wholesale decimation of trilobite diversity (along with many other groups), took place following the appearance of the first forests on land. Intuitively, it would seem highly unlikely for there to have been any correlation whatsoever between these two events. How could the growth of forests impact in any way on the evolutionary well-being of a group of tiny marine arthropods? Yet the rocks whose deposition as sediments straddled this great mass extinction event contain evidence that events on land were, in fact, directly involved in extinctions in the oceans.

Siltstones are generally not the most attractive of rocks. Yet there are some, sporting a variety of stylish colours, that contain within them evidence of Earth's first forests. Near the small town of Cairo in New York State lies an abandoned quarry. As well as being a source of building stone, the site has for many years yielded abundant fossils, especially plants and fishes. One part of the quarry floor exposes an extensive layer of siltstone, about 3,000 square metres (3,590 sq. yd), which represents the upper part of an ancient soil horizon. Within it, exposed to the modern-day elements, are a spectacular array of fossilized tree roots. These are roots that, 385 million years ago, nourished Earth's earliest known forests.[7]

The siltstones in which trees once grew are a kaleidoscope of colour, from shades of red and mottled bluish grey to yellow and brown, to grey, where the roots are best preserved. This multicoloured fossil soil is overlain by a thin layer of greenish siltstone that contains the scattered remains of jawless agnathan and armoured placoderm fishes, along with primitive sharks. The siltstones, the roots and the fossils all tell a story of a forest of large trees inundated during a flood event that killed the fishes. Trapped by the trees, they perished in a sea of mud and silt. All that now remains of this once-vibrant forest is its intricate root system.

This forest consisted of at least three types of trees. One, possibly related to the lycopods (clubmosses). Another, cladoxylopsids, leafless trees up to 10 metres (33 ft) tall with celery-like outgrowths and shallow roots. The third, and most dominant, with the deepest, most extensive root systems that extended at least 1.6 metres (5¼ ft) into the soil, was a tree called *Archaeopteris*. Widespread in Middle to Late Devonian times, it is a member of a group of trees called progymnosperms, which were thought to have been ancestral to modern-day conifers and cycads. More than 20 metres (65 ft) in height, and with

Extensive fossilized root system of 385-million-year-old Middle Devonian progymnosperm tree *Archaeopteris*. Preserved in an abandoned quarry at the town of Cairo, New York. Scale bar (near centre) is 1 metre.

a root spread of up to 11 metres in the quarry, *Archaeopteris* has been described as having looked like a top-heavy Christmas tree, its conifer-esque trunk sprouting large, fern-like leaves. What these three types of trees have in common is that none produces seeds; all reproduced by shedding spores.

The importance of *Archaeopteris* and its roots to Earth's early environment cannot be overstated. The development of deeper-rooted plants during the later part of the Devonian saw the establishment of thicker, more complex soils. Where once soils were a thin veneer of broken rocks a few centimetres deep, roots played a major role in breaking up surface rocks, facilitating the addition of nutrients to the mix. Acids produced by mycorrhizal fungi, along with bacterial decomposition of plant material, produced organic acids, causing chemical weathering of the land surface, leading to the formation of fine clays. One of the most significant effects, apart from adding more nutrients, was the role of soils in drawing down carbon dioxide from the atmosphere. The deeper the soils, the greater the drawdown. This caused a significant reduction in levels of atmospheric carbon dioxide and consequent reduction in atmospheric temperatures. In late Silurian times, about 400 million years ago, atmospheric CO_2 levels were between ten and eighteen times higher than today, but by the Late Devonian this had reduced to between two and five times higher. While the increase in nutrients in these new, deeper soils facilitated the growth of the forests, some, inevitably, found their way into rivers, and thus into the marine environment. But another factor lent a helping hand in delivering these nutrients into the oceans: fire.

* * *

AN UNASSUMING OUTCROP of rocks can be found in the Ardennes in Belgium, near the village of Sinsin. Preserved within is the key to explaining, perhaps, the cause of the extinctions of so many marine species of animals 372 million years ago. Like the Languedoc and Kimberley rocks, the outcrop conveniently straddles the marine world before and after the Late Devonian mass extinction. An ocean of plenty followed by one decimated by the vagaries of forests, fires and rivers. In the pre-extinction sea, sediments deposited were primarily carbonates – represented now by a stratigraphic sequence of limestones. Immediately following the extinctions, sedimentation changed abruptly to dark grey siliceous muds. Like the limestones, these had been deposited in shallow seas off the southern end of the other great supercontinent, Laurasia. The sedimentary rocks now outcropping in the wooded hills of the Ardennes contain two lines of evidence that point to intense wildfires raging on land around the time of the mass extinctions. The first is the presence in the mudstones of fossilized charcoal derived from the combustion of land plants and washed into the sea with the mud.

Wildfires have ravaged this planet since plants first took root on land. Fossil charcoal makes its first appearance in the rock record in 415-million-year-old grey siltstones in Shropshire and is the burnt remains of some of the earliest known land plants.[8] Despite the vegetation being small – less than a metre high – the fires that burnt it would have been sufficient to produce charcoal. The charcoalification resulted in remarkable preservation of cellular structures of the plants, allowing their affinities to be determined. As forests developed in size and floral complexity during the Devonian Period, so did the frequency and intensity of wildfires. The charcoal was washed into the shallow seas along the margins of Laurasia and Gondwana, accumulating in the sediment. The rocks from Sinsin in the Ardennes show a particularly intense increase in charcoal at the level of the maximum extinction event at about 372 million years.[9] While it could be argued that more charcoal derived from land plants in marine sediments might just reflect increased river flow into the shallow seas, there is a second line of evidence, trapped in the mudstones, that also tells of unremitting wildfires – of a land on fire – fossilized polycyclic aromatic hydrocarbons, organic chemicals preserved in rocks for hundreds of millions of years. Where limestone gives way to mudstone and faunal biodiversity plummets, a distinctive suite of these aromatic hydrocarbons shows a pronounced spike in concentration in the rock.

One group of these compounds only forms from the high-temperature combustion of terrestrial plant material. These include the organic compounds coronene, benzo(ghi)perylene and benzo(e)pyrene. Their presence in marine sediments is a result of terrestrial wildfires in the *Archaeopteris*-dominated forests. Earlier in the Devonian, the atmospheric level of oxygen was about 13 per cent (it is 20.95 per cent today). By the Late Devonian this is thought to have risen to about 17 per cent and exacerbated the extent and intensity of the wildfires. Naturally, one of the consequences of wildfires is the destruction of the vegetation. As a result the soil is more susceptible to erosion in the aftermath of the fires, transporting increasing amounts of sediment and nutrients into the shallow marine ecosystem. The soils in the later Devonian being much deeper than in the earlier Devonian, due to the increase in size of many of the dominant trees, would also have led to greater amounts of soil, charcoal and aromatic hydrocarbons being washed into the sea. Further evidence for heightened soil erosion comes from the enhanced presence of another aromatic hydrocarbon, dibenzofuran, which is derived from the breakdown of terrestrial lichen and woody plants. Such material naturally accumulates in the soil. Its preservation in marine sediments is a key marker for substantial soil erosion.

In summary, the mechanisms that in all likelihood led to the Late Devonian mass extinction initially involved sediments derived from the hinterland being washed into lowland near the coast. This was trapped by vegetation in the archaeopterid forests and contributed to soil formation. During the period of lower atmospheric oxygen, there was a build-up of woody fuel, and as oxygen levels rose in the Late Devonian, combustion of this material became more frequent. These fires destroyed much of the plant root structure and the subsequent soil erosion released nutrient-rich soil into rivers, which then washed into the sea. The increase in nutrients, especially phosphorus, spurred proliferation of algal blooms which depleted the water

Impression of the trunk of an undescribed Late Devonian lycopod (clubmoss) from near Kununurra, Western Australia. Typical element of the earliest forests.

of oxygen.[10] Moreover, increased sediment input disrupted filter-feeding organisms. It also totally disrupted those ecosystems, especially reef ecosystems, adapted to low-nutrient conditions, resulting in the extinction of large suites of organisms adapted to such environments. So, what of the trilobites? The scutelluids, odontopleurids and harpetids were unable to evolve quickly enough to survive this drastically changed environment, and all became extinct. Phacopids responded by becoming larger and developing more lenses in their eyes, much like forms from early in their evolutionary history. But a similar nutrient enrichment of the seas took place at the end of the Devonian, and this was too much, even for this group, which also became extinct.[11]

The scenario that played out during Late Devonian times – of soil erosion on land, leading to masses of fine sediments enriched with excess nutrients being released into a shallow marine reef environment resulting in a mass extinction of marine life – is the story told by the rocks. Yet we do not have to look too far to see this being repeated once again – especially if you happen to live near the eastern coast of Australia. Here, the Great Barrier Reef is in danger of replaying the extinctions seen in the Late Devonian rocks. Soil erosion in the reef's hinterland is rampant, produced now by land clearing of vegetation for agricultural purposes rather than by wild-fires, but having the same effect of facilitating soil erosion during periods of high rainfall. And as the fine sediment washes into the coastal reef environment, the sea becomes increasingly turbid. Rivers also wash in excess nutrients – not those produced by forest fires, but by the agricultural practice of application of chemical fertilizers, contributing high concentrations of nitrogen and phosphorus into the soil. Washed into the ocean, these nutrients disrupt the balanced low-nutrient status of the reef ecosystem. Sugarcane farming alone has been shown to contribute 78 per cent of the dissolved inorganic nitrogen into the Great Barrier Reef system.[12]

Time, perhaps, to listen a little more carefully to what rocks have to tell us.

Death of a Forest

When not exercising his new-found interest in collecting rocks, John Woodward is indulging his other passion: plants. It is 1691, and just three years since his rock collection began to grow. But now he is investigating how plants grow. The prevailing wisdom at the time is that they do so simply by taking up water. The more water they suck up, it is thought, the larger they grow. As Woodward was to write some eight years later,

> The great restorer of Philosophy in this last Age, my Lord
> Bacon is of opinion, That for Nourishment of Vegetables, the
> Water is almost all in all: and that the Earth doth but keep
> the Plant upright, and save it from over heat, and over cold.
> Others ... assert Water to be the only Principle or Ingredient
> of all natural things.[13]

Woodward, however, has other ideas, and, being a good disciple of
the philosophy of empiricism of Francis Bacon (1561–1626), sets
out a series of controlled experiments that continue well into 1692,
using the plants spearmint and nightshade. He grows them in waters
containing different amounts of 'terrestrial Matter': spring water,
rainwater and Thames water. In another experiment he grows the
plants in Hyde Park conduit water; then to the same water he adds
either dissolved 'Garden Mould', nitre or 'Common Garden Earth'.
Last, some plants are grown in distilled Hyde Park conduit water
alone.

Meticulously he weighs the plants as he starts his experiments.
Following a set period of growth, he weighs them again. The results
are quite clear. It is not the water that has contributed to the increase
in weight of his plants. It is the *terrestrial Matter* (that is, nutrients)
which plays the major role. Observing that the plants are continually
taking up water, which acts purely as a conduit for the nutrients, he
surmises that water passes out of the plant leaves through little pores
that we now call stomata: 'The much greatest part of the Fluid Mass
that is thus drawn off and convey'd into the Plants, does not settle
or abide there: but passes through the Pores of them, and exhales up
into the Atmosphere.'[14]

Woodward is the first to discover that plants transpire. He also
acknowledges the importance of sunlight on plant growth, but thinks,

not unreasonably, that it is the temperature, rather than the light, that promotes growth – more heat, more growth. It is left to Stephen Hales thirty years later to see the light (as it were) and suggest that sunlight itself plays a critical role in plant growth. Hales also, in his *Vegetable Staticks* of 1726, calculated the substantial amount of moisture that trees added to the atmosphere.[15] But, crucially, Woodward observes that the hotter the environment in which plants grow, not only is growth more pronounced, but there is greater transpiration. Plants; the Sun; heat; water. Woodward recognizes that plant transpiration and rainfall must be linked:

> This so continual an Emission and Detachment of Water, in so great Plenty from the Parts of Plants, affords us a manifest reason why Countries that abound with Trees and the large Vegetable especially, should be very obnoxious to Damps, great Humidity in the Air, and more frequent Rains, than other that are more open and free. The great moisture in the Air, was a mighty inconvenience and annoyance to those who first settled in America; which at that time was much overgrown with Woods and Groves. But as these were burnt and destroyed, to make way for Habitation and Culture of the Earth, the Air mended and cleared up space: changing into a Temper much more dry and serene than before.[16]

Woodward's conclusions were clear: remove forests and rainfall declines.

* * *

A ROCK IN THE HAND can only tell you so much about the past. It is, in many ways, often just the faintest blink of a time long passed. However, take a rocky outcrop on an exposed coastal cliff: it contains far more complex stories beckoning to be unravelled. It contains the luxury of time – stories of ancient worlds, of changing environments – the fauna, flora, climate, topography. To read such a cliff it is necessary, as with a good book, to start at the beginning. In this case, at the bottom of the outcrop. (Imagine trying to read a book that is lying face-down in the dirt. You would read it from bottom to top.) Passing up the rock sequence is to read such a geological book. Sometimes, like chalk cliffs, much the same story is repeated many times – chalk and flint; chalk and flint; on and on. But there are some outcrops of rocks that can tell stories of startling events in the history of Earth: the rocks, varying in colour, texture, chemistry; the fossils they contain, from the microscopic to the macroscopic, changing, sometimes subtly, other times dramatically. This is the beauty of stratigraphy. All elements contained within the rocks are parts of the memories of environments that have come and gone – and often in quite a dramatic fashion.

Clinging to the eastern coast of Australia at Frazer Beach, some 100 kilometres (62 mi.) north of Sydney, is a 16-metre-high (52 ft) cliff. The rocks of which it is composed vary between black and shades of grey.[17] Like the outcrops in the Ardennes, this cliff also straddles a mass extinction boundary – the greatest the world has ever experienced. Woodward would have been interested by this one, as it seems to show that what he described as occurring in seventeenth-century America might have affected much older forests – 252 million years older to be reasonably precise – the loss of vegetation replaced by a much drier world. It is a story that starts at the bottom of the cliff, in black rocks that were once alive, for to stand here is to set foot in an ancient forest that thrived before the mass extinction. The black rock is coal – the combustible fossilized remnants of ancient forests. Reaching barely to knee height, this thin layer of coal is the last of many found throughout more than a kilometre of sediments that form the underworld in this part of Australia.

Like all coals it started life as rotting vegetation – leaves, branches, seeds, trunks – the fallen remains of once-extensive and long-lived forests. Settling in mires – wet, swampy environments – the vegetation slowly decomposed and as it accumulated, became compressed into peat. Periodic inundations, either by river systems or by rising sea levels, deposited sediments over the peat beds, and steadily they were buried. With this came increasing pressure and heat, and over millions of years the peat was transformed into coal, first brown coal, or lignite, then eventually into more carbon-rich black coal: anthracite. While coal deposits in the Northern Hemisphere are dominated by Carboniferous coals, formed from forests of huge lycopods (club-mosses), equisetids (horsetails) and tree ferns that lived 325 to 300 million years ago, those in eastern Australia, like other parts of the Southern Hemisphere, formed more recently, in Permian times, between 300 and 250 million years ago. And these southern forests were quite distinctive in their botanical make-up.

In 1828, some fossil leaves from Permian rocks of India were described by the French botanist Adolphe Brongniart. He gave them the name *Glossopteris*. These leaves were found preserved in coal and in the intervening repeated cycles of silts and muds. During the rest of the nineteenth century, similar leaves came to light in coal-bearing sedimentary rocks in Australia, South Africa and South America. In part on the basis of the distribution of these plants on such widely spaced southern continents, geologists from the Geological Survey of India proposed that during Carboniferous and Permian times a great southern continent had existed. In 1885, Austrian geologist Edward Suess gave it the name Gondwana-Land (now just Gondwana). But there was one piece missing in the Gondwanan jigsaw puzzle: Antarctica.

* * *

IT IS 8 FEBRUARY 1912, and it is bitterly cold. Robert Scott, with his four companions, Edward Wilson, Lawrence Oates, Henry Bowers and Edgar Evans, are trekking back from the South Pole. Three weeks earlier, on reaching the pole, they were devastated to find a Norwegian flag being buffeted by the wind. Roald Amundsen had beaten them to it. Leaving the Polar Plateau, they track down the Beardmore Glacier. In an attempt to escape the incessant wind they head to a moraine below Mount Buckley. 'A lot could be written on the delight of setting foot on rock after 14 weeks of snow and ice,' Scott writes in his diary. 'It is like going ashore after a sea voyage.' Scott decides that they should camp on the moraine 'and spend the rest of the day geologising'. He found it

> extremely interesting. We found ourselves under perpen-
> dicular cliffs of Beacon sandstone, weathering rapidly and
> carrying veritable coal seams. From the last Wilson, with his
> sharp eyes, has picked several plant impressions, the last a
> piece of coal with beautifully traced leaves in layers, also some
> excellently preserved impressions of thick stems, showing
> cellular structure.[18]

Despite weighing about 16 kilograms (35 lb), Scott is determined to take these and other rocks collected from the moraine with them. For the next six weeks, as they struggle to return to base camp, losing first

Evans and then Oates on the way, the rocks are their quiet companions. However, destiny deprives them of the chance to take the rocks and *Glossopteris* fossils back to England themselves. Nine months after they collected the rocks, their frozen bodies are discovered, and with them the rocks containing the fossil *Glossopteris* leaves. Their geological legacy is immense, for it showed that, as on the other southern continents, *Glossopteris* forests once thrived on Antarctica. Thanks to Scott and his colleagues, the jigsaw is complete.

* * *

A WANDER (or a wade, more likely) through a swampy bald cypress forest in the southeast United States is probably the closest it is possible to get today to experiencing life in a Permian glossopterid forest. The variety of fossil plants preserved in Permian rocks that also contain *Glossopteris* is somewhat limited, suggesting that this rather strange tree dominated these Gondwana forests. Growing to a height of about 30 metres (98 ft), individual trees were slender and conical, sporting tongue-shaped, veined leaves that hung vertically like icicles from the branches. This was probably an efficient way of dealing with the low-angle sunlight that bathed the trees in the high latitudes in which the forests grew. Their fossilized root systems show that beneath the ground the trees had specialized vacuoles which would have facilitated aeration of the roots in waterlogged soils.

Given the high density of leaves found fossilized in rocks, it is likely that the trees were deciduous. Such a habit might have arisen from their growth at very high latitudes. The fossils found on Scott's polar expedition were found at a latitude of about 85°s. Even in Permian times, this part of Gondwana would have lain at a similar latitude. So, what enabled forests to live there then and not now? Atmospherically, CO_2 levels at that time were much higher than today

Leaves of the late Permian tree
Glossopteris from Dunedoo, New
South Wales.

and, combined with their southern position on the huge continent of Pangaea that extended from pole to pole, meant that even at the poles, temperatures were warm and ice-free. Yet the trees still needed growth strategies to cope with long periods of darkness. This is why *Glossopteris* probably adapted to a twilight world by being deciduous, shedding its leaves over a short period annually. Fossil leaves tend to be found in concentrations in distinct layers in the rock. During the dark, but warm, winter months, the trees would have not needed to photosynthesize, so there was no point in hanging on to their leaves. Winter dormancy is also shown by pronounced seasonal patterns in growth increments in leaf spacing along shoots. Likewise, fossil *Glossopteris* wood has very well-defined growth rings, indicating seasonal growth.

After they had dominated the southern continent for millions of years, the demise of *Glossopteris* forests was as sudden as it was brutal. As with other coal-bearing rock sequences formed at the end of the Permian, the cliff at Frazer Beach shows a dramatic change in rock type. Lying on top of the thin, basal coal layer is a layer of broken organic fragments 20 centimetres (8 in.) thick. These are essentially angular fragments of coal and charcoal. This, the post-extinction 'dead zone', marks the greatest extinction of life globally. It has been estimated that at least 90 per cent of all living species became extinct. Travelling through time, the cliff sequence passes up into several metres of dark grey mudstones and siltstones. Within these are also fine charcoal particles, degraded woody debris and fungal remains.[19]

With the abrupt die-off of vegetation, the exposed woody biomass and dried peat surfaces would have been prone to wildfires. Evidence for this is found in a spike in concentration of charcoal in the sedimentary 'dead zone' at Frazer Beach, immediately above the coal layer. A similar event has been documented in latest Permian tropical peatlands in China, where there was an intensification of wildfires immediately after the end-Permian extinction event.[20] Low percentages of charcoal in sediments prior to the event are replaced by a nearly sixfold increase in the post-extinction sediments. With increased burning, combined with abruptly altered precipitation and land surface temperatures, vegetation dieback was enhanced, as was seedling recruitment.[21] It is likely that increased frequency of wildfires may have played a crucial role in the collapse of continental ecosystems during the end-Permian extinction event, even in areas populated by fire-tolerant wetland vegetation. As well as engendering massive ecological stress for terrestrial life, wildfires probably served as positive feedback for greenhouse-gas-driven warming, and a mechanism that catalysed toxic microbial blooms in regional aquatic ecosystems.[22]

With the loss of the forests would have come rising water tables. The muds and silts deposited on top of the coal are interpreted as sediments deposited in lakes following the extinction of the forests. Peat formation ceased, and extensive, ephemeral ponds resulted from cessation of organic accumulation following the extinction of the *Glossopteris*-dominated mire forests.[23] Research on modern wetland systems has shown that complete deforestation can result in a 15 per cent increase in volumes of water, leading to ephemeral lake systems being converted into permanent wetlands.[24]

While the forests had been growing during Permian times, climatic conditions were consistently warm and moist, despite being influenced by a distinct photoperiod owing to their high latitude setting. The Frazer Beach rocks high in the cliff section above the extinction boundary contain the fossilized remains of a flora that began to appear during earliest Triassic times. It is dominated by gymnosperms (the group that contains conifers) with small leaves, thick cuticles and strongly protected stomata (to prevent excess water loss).[25] These are all features typical of sclerophyllous plants

– forms adapted to low-nutrient soils and with the ability to withstand seasonal dry conditions. The interpretation from the sediments and fossils preserved in the rocks at Frazer Beach point to a post-extinction change from a warm, humid climate to a dryer one, with seasonally intensified moisture deficiency.[26] This raises the question of whether the climate change *caused* the extinction, or whether the loss of the forests was itself a prime factor in altering the climate. Many explanations have been proposed for what caused the end-Permian mass extinction. The one most favoured today is the effect of massive volcanic eruptions that occurred at this time, especially in the so-called Siberian Traps. The eruption of extensive flood basalts over an area of roughly 7 million square kilometres (nearly 3 million sq. mi.) would have released huge quantities of gases, especially SO_2 and CO_2, into the atmosphere, heating it and acidifying the oceans.

John Woodward was by no means the first to propose a link between deforestation and climate change. The ancient Greek philosopher Theophrastus (*c.* 371–*c.* 287 BC), in his writings on plants in *Historia Plantarum*, had already observed the relationship. This idea gathered strength during the Renaissance.[27] Christopher Columbus, for one, feared that deforestation in the Canary Islands would lead to a significant decline in rainfall. In seventeenth-century England, Francis Bacon and Edmond Halley theorized about the links between vegetation, rainfall and the hydrological cycle. And all this before John Woodward's musings.

By the mid-eighteenth century, linking deforestation to rainfall reduction (or desiccation theory, as it became known) formed the

Cliff section through the Permian–Triassic mass extinction event at Frazer Beach, New South Wales, showing Permian *Glossopteris* coal forest abruptly transitioning to a wildfire/fungal event preserved in Triassic shales. Palaeobotanist Steve McLoughlin climbing though Triassic sediments for scale.

platform for the development of British and French colonial forest protection systems, particularly as they affected forestry policy in the West Indies.[28] Although interest in desiccation theory dwindled in the twentieth century, it has been rekindled in recent times as concern about the impact of human activity on climate change has gained more traction. One recent case exemplifies the negative effect of large-scale removal of vegetation over a short time period, perhaps analogous to the rapid loss of the Permian *Glossopteris* forests.[29]

Since the arrival of the first permanent Europeans settlers in Australia, a little over two hundred years ago, removal of natural vegetation has been undertaken on an almost industrial scale. For instance, in the southwest region of Western Australia, which covers an area of some 196,000 square kilometres (75,000 sq. mi.), it has been estimated that, mostly through the twentieth century, 80 per cent of the natural vegetation has been removed to make way predominantly for wheat and sheep. The consequence of this has been waterlogging of the land, as the water table inevitably rose, but also a concomitant reduction in regional rainfall. The region has suffered from a steady decline in rainfall attributed to 'climate change' over the past forty years. However, it has been estimated that 55 to 62 per cent of this decline can be attributed directly to the removal of this natural vegetation cover.[30] If such a relatively local change can have a profound effect on rainfall, then the transition from coal to mudstones and siltstone at Frazer Beach poses the question of whether the global loss of an entire forest ecosystem over a very short period, geologically speaking, could have played a part in the transition to a much more arid climate following the end-Permian extinction event.

* * *

ON THE SURFACE, the study of rocks would hardly seem to be one of the most scintillating facets of geology. Take palaeontology: it boasts a stunning cornucopia of creatures that have inhabited this planet for hundreds of millions of years, from feathered dinosaurs to spiders trapped in amber, to trilobites sporting an array of armour guaranteed to have made a medieval knight jealous. Or mineralogy: glistening with gold, diamonds and countless varieties of spectacular crystals. Rocks, though? Endless piles of mudstones, siltstones, sandstones and limestones. Yet out of these often seemingly dull rocks, like the black coal and grey mudstone at Frazer Beach or the bland limestone and mudstone in the Ardennes, come a welter of stories, telling of evolution and extinction in the biological world; of changing climates; of the rise and fall of great forests; and of a planet ravaged by wildfires.

Ignore rocks at your peril.

EPILOGUE

Vast swathes of the underworld are hidden from view – by cities and villages, forests and fields, roads and meadows. So many past worlds are lost to us. But the underworld can be glimpsed, not only where the sea gnaws at its edges or in the flanks of rivers, but in the pits and quarries that humans have dug to extract rocks that they find useful. Yet quarries are transient features: once their sell-by dates are reached and they are no longer productive, they become perceived as societal encumbrances to be removed at all costs. Eyesores in the landscape. They are filled with garbage, then have buildings or fields planted on them. Or they are drowned by water and transformed into aquatic playgrounds. All activities to ensure the 'unsightly' underworld is banished from our sight. Yet, by 'tidying up' the landscape, so much of our ability to understand how the world has evolved and come to be what it is today is lost.

However, once in a while the serendipitous discovery of an overgrown quarry can reveal why some of these windows into former worlds need to be kept. As I finished writing the last chapter of this book, a news report appeared regarding the discovery of a long-forgotten quarry by a pair of 'amateur' palaeontologists. What they found was a meadow hidden in woodland. But this was not a meadow full of wildflowers. Rather, it was an echinoderm meadow that 167 million years ago teemed with life. Now it is just a bed of limestone covered by a layer of mud. Yet within the limestone, on the floor of the quarry, lie the fossilized remains of the seabed frozen in time, not by ice but by a slurry of mud that catastrophically smothered a vibrant, living community of sea urchins, starfishes, stalked crinoids and feather stars. Thousands upon thousands of exquisitely preserved specimens, the likes of which have never previously been found in Jurassic rocks in Britain, snuffed out in an instant by a storm event or an earthquake, dumping mud over a once-thriving echinoderm community.

While not all abandoned, unfilled quarries yield such spectacular insights into past worlds, they all do have some sort of story to tell – of ancient environments, of changing climates and of how organisms have contributed to making so many types of rocks. Sometimes, just sometimes, it would seem worthwhile to preserve a few of these intriguing portals through which we can unearth the underworld.

REFERENCES

1 EARTH'S DIRTY LITTLE SECRETS

1 Revd William Cole's Collections, vol. XXXIII, British Library, Add MS 5834
F. 156; R. McCormmach, *Weighing the World: The Reverend John Michell of
Thornhill* (Berlin/Heidelberg, 2012), p. 40.
2 McCormmach, *Weighing the World*.
3 John Michell, 'Conjectures Concerning the Cause, and Observations Upon
the Phaenomena of Earthquakes . . .', *Philosophical Transactions*, LI (1759),
p. 585.
4 Ibid., p. 582.
5 William H. Fitton, 'Notes on the History of English Geology', *Philosophical
Magazine*, I (1832), pp. 268–75.
6 Oxford University Museum of Natural History, ed., *Strata: William Smith's
Geological Maps* (London, 2020).
7 John Woodward, *An Attempt Towards a Natural History of the Fossils of
England; in a Catalogue of the English Fossils in the Collection of J. Woodward,
MD Containing a Description and Historical Account of Each; with Observations
and Experiments, Made in Order to Discover, as Well the Origin and Nature
of Them, as Their Medicinal, Mechanical, and Other Uses. Tome 1, Part 1*
(London, 1729), p. 45.
8 Ibid.
9 Ibid., p. 46.
10 Ibid., p. x.
11 Ibid., pp. xiii–xiv.
12 John Woodward, *A Catalogue of the Additional English Native Fossils, in the
Collection of J. Woodward MD Tome II. A Catalogue of the Second Addition of
English Native Fossils* (London, 1728), p. 63.
13 Ibid., p. 99.
14 V. Paul Wright et al., 'The Paleohydrology of Lower Cretaceous Seasonal
Wetlands, Isle of Wight, Southern England', *Journal of Sedimentary Research*,
LXX/3 (2000), pp. 619–32.
15 John Woodward, *An Addition to the Catalogue of the Foreign Native Fossils, in
the Collection of J. Woodward MD* (London, 1728), p. 20.
16 John van de Bemde (*c.* 1655–*c.* 1726).

17 Thomas W. W. Hearing et al., 'Quantitative Comparison of Geological Data and Model Simulations Constrains Early Cambrian Geography and Climate', *Nature Communications*, XII/1 (2021), accessed at www.nature.com.

18 Boriana Kalderon-Asael et al., 'A Lithium-Isotope Perspective on the Evolution of Carbon and Silicon Cycles', *Nature*, DXCV/7867 (2021), pp. 395–8.

19 Alan Buis, 'Milankovitch (Orbital) Cycles and Their Role in Earth's Climate' (2020), https://climate.nasa.gov.

20 Stephen R. Meyers and A. Malinverno, 'Proterozoic Milankovitch Cycles and the History of the Solar System', *Proceedings of the National Academy of Sciences of the United States of America*, CXV/25 (2018), pp. 6363–8.

21 Ken McNamara, *Dragons' Teeth and Thunderstones: The Quest for the Meaning of Fossils* (London, 2020).

2 INTO THE LIGHT

1 Kathleen Grey and S. Awramik, 'Handbook for the Study and Description of Microbialites', *Bulletin of the Geological Survey of Western Australia*, CXLVII (2020), pp. 1–277.

2 Kathleen Grey and N. J. Planavsky, 'Microbialites of Lake Thetis, Cervantes, Western Australia – A Field Guide', *Records of the Geological Survey of Western Australia*, XI (2009), pp. 1–28.

3 R. Pamela Reid et al., 'The Role of Microbes in Accretion, Lamination and Early Lithification of Modern Marine Stromatolites', *Nature*, CDVI/6799 (2000), pp. 989–92.

4 Zhongwu Lan et al., 'Evidence for Microbes in Early Neoproterozoic Stromatolites', *Sedimentary Geology*, CCCXCVIII (2020).

5 Roger Buick and J. Dunlop, 'Evaporitic Sediments of Early Archaean Age from the Warrawoona Group, North Pole, Western Australia', *Sedimentology*, XXXVII (1990), pp. 247–77.

6 Malcolm R. Walter, R. Buick and J. Dunlop, 'Stromatolites 3,400–3,500 Myr Old from the North Pole Area, Western Australia', *Nature*, CCLXXXIV/5755 (1980), pp. 443–5.

7 Abigail C. Allwood et al., 'Stromatolite Reef from the Early Archaean Era of Australia', *Nature*, CDXLI (2006), pp. 714–18.

8 Raphael J. Baumgartner et al., 'Nano-Porous Pyrite and Organic Matter in 3.5-Billion-Year-Old Stromatolites Record Primordial Life', *Geology*, CLVII/11 (2019), accessed at https://pubs.geoscienceworld.org.

9 Ibid.

10 William Dampier, *A Voyage to New Holland, &C. in the Year, 1699* (London, 1703), vol. III, p. 124.

11 Phillip E. Playford et al., 'The Geology of Shark Bay', *Bulletin of the Geological Survey of Western Australia*, CXLVI (2013), p. 146.

12 Ibid., pp. 116–46.

13 Erica P. Suosaari et al., 'New Multi-Scale Perspectives on the Stromatolites of Shark Bay, Western Australia', *Scientific Reports*, VI (2016), pp. 1–13.

14 Stanley M. Awramik and H. P. Buchheim, 'A Giant, Late Archean Lake System: The Meentheena Member (Tumbiana Formation; Fortescue Group), Western Australia', *Precambrian Research*, CLXXIV (2009), pp. 215–40.

15 Ibid.

16 Ken McNamara, *Stromatolites* (Perth, 2009).

17 Maurice Tucker, 'The Precambrian–Cambrian Boundary: Seawater Chemistry, Ocean Circulation and Nutrient Supply in Metazoan Evolution, Extinction and Biomineralization', *Journal of the Geological Society*, CIVIX (1992), pp. 655–68.

18 William Buckland and H. De La Beche, 'On the Geology of the Neighbourhood of Weymouth and the Adjacent Parts of the Coast of Dorset', *Transactions of the Geological Society of London*, IV (1835), pp. 1–46.

19 Arnaud Gallois, Dan Bosence and Peter M. Burgess, 'Brackish to Hypersaline Facies in Lacustrine Carbonates: Purbeck Limestone Group, Upper Jurassic–Lower Cretaceous, Wessex Basin, Dorset, UK', *Facies*, LXIV (2018).

3 ROCK OF AGES

1 John P. Grotzinger and N. P. James, 'Precambrian Carbonates: Evolution of Understanding', *Carbonate Sedimentation and Diagenesis in the Evolving Precambrian World*, SEPM *Special Publication*, 67 (2000), pp. 3–20.

2 Dawn Y. Sumner, 'Facies, Paleogeography, and Carbonate Precipitation in the Archean (2520 Ma) Campbellrand-Malmani Carbonate Platform, Transvaal Supergroup, South Africa', PhD diss., Massachusetts Institute of Technology, Cambridge, MA (1995), p. 514.

3 Grotzinger and James, 'Precambrian Carbonates', p. 13.

4 John Woodward, *An Attempt Towards a Natural History of the Fossils of England; in a Catalogue of the English Fossils in the Collection of J. Woodward, MD Containing a Description and Historical Account of Each; with Observations and Experiments, Made in Order to Discover, as Well the Origin and Nature of Them, as Their Medicinal, Mechanical, and Other Uses. Tome 1, Part 2* (London, 1729), p. 86.

5 Robert Hooke, *Micrographia; or, Some Physiological Descriptions of Minute Bodies Made by Magnifying Glasses, with Observations and Inquiries Thereupon* (London, 1665).

6 Ibid., p. 93.

7 Ibid., p. 94.

8 Ibid., p. 95.

9   David T. Flannery et al., 'Microbially Influenced Formation of Neoarchean Ooids', *Geobiology*, XVII/2 (2019), pp. 151–60.

10 Mara R. Diaz et al., 'Microbially Mediated Organomineralization in the Formation of Ooids', *Geology*, XLV/9 (2017), pp. 771–4.

11 Kenneth J. McNamara, 'Earth and Life – Origins of Phanerozoic Diversity', *Australian Journal of Earth Sciences*, LV (2008), pp. 1023–36.

12 Nicholas J. Butterfield, 'Plankton Ecology and the Proterozoic–Phanerozoic Transition', *Paleobiology*, XXIII (1997), pp. 247–62.

13 Simon Conway Morris, 'The Community Structure of the Middle Cambrian Phyllopod Bed (Burgess Shale)', *Palaeontology*, XXIX (1986), pp. 423–67.

14 Gonzalo Vidal and M. Moczydłowska-Vidal, 'Biodiversity, Speciation, and Extinction Trends of Proterozoic and Cambrian Phytoplankton', *Paleobiology*, XXIII/2 (1997), pp. 230–46.

15 Butterfield, 'Plankton Ecology and the Proterozoic–Phanerozoic Transition'.

16 Graham A. Logan et al., 'Terminal Proterozoic Reorganization of Biogeochemical Cycles', *Nature*, CCCLXXVI (1995), pp. 53–6.

17 Rafie Shinaq and K. Bandel, 'Microfacies of Cambrian Limestones in Jordan', *Facies*, XXVII (1992), pp. 52–7.

18 Thomas Servais and D. Harper, 'The Great Ordovician Biodiversification Event (GOBE): Definition, Concept and Duration', *Lethaia*, LI/2 (2018), pp. 151–64.

19 Mark T. Gibbs et al., 'An Atmospheric pCO$_2$ Threshold for Glaciation in the Late Ordovician', *Geology*, XXV/2 (1997), pp. 447–50.

20 Servais and Harper, 'Great Ordovician Biodiversification Event (GOBE)'.

21 Phillip E. Playford et al., 'Devonian Reef Complexes of the Canning Basin, Western Australia', *Bulletin of the Geological Survey of Western Australia*, CXLV (2009), pp. 1–444.

22 Rachel Wood, 'Palaeoecology of a Late Devonian Back Reef Canning Basin, Western Australia', *Palaeontology*, XLIII/4 (2000), pp. 671–703.

23 Hengchao Xu et al., 'Precipitation of Calcium Carbonate Mineral Induced by Viral Lysis of Cyanobacteria: Evidence from Laboratory Experiments', *Biogeosciences*, XVI (2019), pp. 949–60.

24 Alois Senefelder, *The Invention of Lithography*, trans. J. W. Muller (New York, 1911), p. 14.

25 Gordon Walkden, *Devonshire Marbles: Their Geology, History and Uses* (London, 2015).

26 Ibid.

27 Bernard Mamet and A. Préat, 'Iron-Bacterial Mediation in Phanerozoic Red Limestones: State of the Art', *Sedimentary Geology*, CLIIIV/3–4 (2006), pp. 147–57.

28 Noel P. James and Y. Bone, 'Palaeoecology of Cool-Water, Subtidal Cycles in Mid-Cenozoic Limestones, Eucla Platform, Southern Australia', *Palaios*, IX (1994), pp. 457–76.

4 AS CHALK IS TO CHEESE

1 John Woodward, *An Attempt Towards a Natural History of the Fossils of England; in a Catalogue of the English Fossils in the Collection of J. Woodward, MD Containing a Description and Historical Account of Each; with Observations and Experiments, Made in Order to Discover, as Well the Origin and Nature of Them, as Their Medicinal, Mechanical, and Other Uses. Tome 1, Part 1* (London, 1729), vol. 1, part 2, pp. 7, 8.

2 Christian G. Ehrenberg, 'Bemerkungen ueber feste mikroskopische anorganische Formen in den erdigen und derben Mineralsen', *Bericht uber die Verhandlungen der Königlich Preussischen Akademie der Wissenschaften, Berlin* (1836), pp. 84–5.

3 Thomas H. Huxley, Appendix to *Captain Dayman's Deep-Sea Soundings in the North Atlantic Ocean Between Ireland and Newfoundland, made in HMS 'Cyclops'*, published by order of the Lords Commissioners of the Admiralty (London, 1858), p. 64.

4 George C. Wallich, 'Remarks on Some Novel Phases of Organic Life, and on the Boring Powers of Minute Annelids, at Great Depths in the Sea', *Annals and Magazine of Natural History*, 3rd series, VIII/45 (1861), pp. 52–8.

5 Henry C. Sorby, 'On the Organic Origin of the So-Called "Crystalloids" of the Chalk', *Annals and Magazine of Natural History*, 3rd series, VIII/45 (1861), pp. 193–200.

6 Thomas H. Huxley, '*On a Piece of Chalk*: A Lecture Given in 1868 to the Workingmen of Norwich', *Macmillan's Magazine*, XVIII (1868), pp. 396–408.

7 John Murray, 'Preliminary Report to Professor Wyville Thomson, FRS, Director of the Civilian Scientific Staff, on Work Done on Board the *Challenger*', *Proceedings of the Royal Society*, XXIV (1876), pp. 471–543.

8 William J. Kennedy and R. E. Garrison, 'Morphology and Genesis of Nodular Chalks and Hardgrounds in the Upper Cretaceous of Southern England', *Sedimentology*, XXII/3 (1975), pp. 311–86.

9 Meredith White et al., 'Coccolith Dissolution within Copepod Guts Affects Fecal Pellet Density and Sinking Rate', *Nature Scientific Reports* (2018), accessed at www.nature.com.

10 Ibid.

11 Andrew Gale et al., 'Orbital Tuning of Cenomanian Marly Chalk Successions: Towards a Milankovitch Time-Scale for the Late Cretaceous', *Philosophical Transactions of the Royal Society, London*, CCCLVII/1757 (1999), pp. 1815–29.

12 Ken McNamara, *Dragons' Teeth and Thunderstones: The Quest for the Meaning of Fossils* (London, 2020).

13 Worthington G. Smith, *Man, the Primeval Savage* (London, 1894).

14 Hans J. P. Zijlstra, 'Sedimentology of the Late Cretaceous and Early Tertiary (Tuffaceous) Chalk of Northwest Europe', *Geologica Ultraiectina, Mededelingen van de Faculteit Aardwetenschappen Universiteit Utrecht*, 119 (1994).

15 William Buckland, 'Description of the Paramoudra, a Singular Fossil Body that Is Found in the Chalk of the North of Ireland; with Some General Observations Upon Flints in Chalk, Tending to Illustrate the History of Their Formation', *Transactions of the Geological Society of London*, 1st series, IV (1817), pp. 413–23.

16 Christopher J. Clayton, 'The Chemical Environment of Flint Formation in Upper Cretaceous Chalks', in *The Scientific Study of Flint and Chert*, ed. G. De C. Sieveking and M. B. Hart (Cambridge, 1986), pp. 43–54.

17 Christopher V. Jeans et al., 'Sulfur Isotope Patterns of Iron Sulfide and Barite Nodules in the Upper Cretaceous Chalk of England and Their Regional Significance in the Origin of Coloured Chalks', *Acta Geologica Polonica*, LXVI/2 (2016), pp. 227–56.

5 A BREATH OF FRESH AIR

1 Not a term usually associated with rocks, but applied to banded ironstones by the late Alex Trendall, former director of the Geological Survey of Western Australia and pioneering geologist on the banded iron formation. He was also a wizard on the squash court. He thrashed me in every one of the many games we played together – each one a masterclass in squash strategy. I never learnt.

2 Priyadarshi Chowdhury et al., 'Magmatic Thickening of Crust in Non-Plate Tectonic Settings Initiated the Subaerial Rise of Earth's First Continents 3.3 to 3.2 Billion Years Ago', *Proceedings of the National Academy of Sciences*, CXVIII/46 (2021).

3 John W. Rogers and M. Santosh, 'Supercontinents in Earth History', *Gondwana Research*, VI/3 (2003), pp. 357–68.

4 David C. Catling and K. J. Zahnle, 'The Archean Atmosphere', *Science Advances*, VI/9 (2020), accessed at www.science.org; J. Farquhar et al., 'Atmospheric Influence of Earth's Earliest Sulfur Cycle', *Science*, CCLXXXIX/5480 (2000), pp. 756–8.

5 Ibid.

6 Yoshiki Kanzaki and T. Murakami, 'Estimates of Atmospheric CO_2 in the Neoarchean–Paleoproterozoic from Paleosols', *Geochimica et Cosmochimica Acta*, CLIX (2015), pp. 190–219.

7 Rebecca C. Payne et al., 'Oxidized Micrometeorites Suggest Either High pCO_2 or Low pN_2 during the Neoarchean', *PNAS*, CXVII/3 (2020), pp. 1360–66.

8 Catling and Zahnle, 'The Archean Atmosphere', p. 7.

9   Victor Von Brunn and D.J.C. Gold, 'Diamictite in the Archaean Pongola
    Sequence of Southern Africa', *Journal of African Earth Sciences (and the
    Middle East)*, XVI (1993), pp. 367–74.
10 Catling and Zahnle, 'The Archean Atmosphere'.
11 Sanjoy Som et al., 'Air Density 2.7 Billion Years Ago Limited to Less Than
    Twice Modern Levels by Fossil Raindrop Imprints', *Nature*, CDLXXXIV/7394
    (2012), pp. 359–62.
12 William S. Cassata and P. R. Renne, 'Fossil Raindrops and Ancient Air',
    Nature, CDLXXXIV/7394 (2012), pp. 322–3.
13 Owen R. Lehmer et al., 'Atmospheric CO_2 Levels from 2.7 Billion Years Ago
    Inferred from Micrometeorite Oxidation', *Sciences Advances*, VI/4 (2020),
    accessed at www.science.org.
14 Som et al., 'Air Density 2.7 Billion Years Ago Limited to Less Than Twice
    Modern Levels by Fossil Raindrop Imprints'.
15 Alec F. Trendall, 'Hailstones Also Fall', *Geoscientist*, published online
    23 August 2012, accessed at www.geolsoc.org.uk.
16 Alec F. Trendall, 'The Significance of Banded Iron Formation (BIF) in the
    Precambrian Stratigraphic Record', *Geoscientist*, X (2000), pp. 4–7.
17 Kurt O. Konhauser et al., 'Could Bacteria Have Formed the Precambrian
    Banded Iron Formations?', *Geology*, XXX (2002), pp. 1079–82.
18 Kaarel Mänd et al., 'Palaeoproterozoic Oxygenated Oceans Following the
    Lomagundi–Jatuli Event', *Nature Geoscience*, XIII/4 (2020), pp. 302–6.
19 Weiqiang Li, B. L. Beard and C. M. Johnson, 'Biologically Recycled
    Continental Iron Is a Major Component in Banded Iron Formations', *PNAS*,
    CXII/27 (2015), pp. 8193–8.
20 Margriet L. Lantink et al., 'Climate Control on Banded Iron Formations
    Linked to Orbital Eccentricity', *Nature Geoscience*, XII/5 (2019), pp. 369–74.
21 Li, Beard and Johnson, 'Biologically Recycled Continental Iron'.
22 Trendall, 'The Significance of Banded Iron Formation (BIF) in the
    Precambrian Stratigraphic Record'.
23 Yi-Liang Li, 'Micro- and Nanobands in Late Archaean and
    Palaeoproterozoic Banded-Iron Formation as Possible Mineral Records
    of Annual and Diurnal Depositions', *Earth and Planetary Science Letters*,
    CCCXCI (2014), pp. 160–70.
24 Gregg Borschmann, Oliver Gordon and Scott Mitchell, 'Rio Tinto Blasting of
    46,000-Year-Old Aboriginal Sites Compared to Islamic State's Destruction in
    Palmyra', www.abc.net.au, 28 May 2020.
25 David J. Flint et al., 'The Importance of Iron Ore to Western Australia's
    Economy', *AusIMM Bulletin*, IX (June 2018), pp. 50–54.
26 Michael Slack et al., 'Aboriginal Settlement during the LGM at Brockman,
    Pilbara Region, Western Australia', *Archaeology in Oceania*, XLV (2009),
    pp. 32–9.

27 Michelle Stanley and Kelly Gudgeon, 'Pilbara Mining Blast Confirmed to Have Destroyed 46,000-y-o Sites of "Staggering" Significance', www.abc.net.eu, 26 May 2020.

28 Steven J. H. Walker et al., 'Kathu Townlands: A High Density Earlier Stone Age Locality in the Interior of South Africa', PLOS One, IX/7 (2014).

29 Francesco Berna et al., 'Microstratigraphic Evidence of In Situ Fire in the Acheulean Strata of Wonderwerk Cave, Northern Cape Province, South Africa', PNAS, CIX/20 (2012), pp. E1215–20.

30 Ibid.

31 L. Wallis et al., 'PXRF Analysis of a Yellow Ochre Quarry and Rock Art Motifs in the Central Pilbara', Journal of the Anthropological Society of South Australia, XL (2016), pp. 134–55.

32 Vicky Winton et al., 'The First Radiometric Pleistocene Dates for Aboriginal Occupation at Weld Range, Inland Mid West, Western Australia', Australian Archaeology, LXXXII/1 (2016), pp. 60–66.

33 John Clarke, 'Two Aboriginal Rock Art Pigments from Western Australia: Their Properties, Use, and Durability', Studies in Conservation, XXI/3 (1976), pp. 134–42.

34 Harry P. Woodward, 'A Geological Reconnaissance of a Portion of the Murchison Goldfield', Bulletin of the Geological Survey of Western Australia, LVII (1914), pp. I–III.

6 RIVERS OF SAND

1 Emrys R. Phillips et al., 'Further Information Relating to "A Geological Perspective on the Stone of Destiny" by N. J. Fortey, E. R. Phillips, A. A. McMillan and M.A.E. Browne', Scottish Journal of Geology, XXXV/2 (1999), pp. 187–8.

2 George Grey, Journals of Two Expeditions of Discovery in North-West and Western Australia, During the Years 1837, 38, and 39 . . . (London, 1841), vol. II.

3 Phillips et al., 'Further Information Relating to "A Geological Perspective on the Stone of Destiny"'.

4 Yawooz A. Kettanah et al., 'Provenance of the Ordovician–Lower Silurian Tumblagooda Sandstone, Western Australia', Australian Journal of Earth Sciences, LXII/7 (2015), pp. 817–30.

5 Simon A. Wilde et al., 'Evidence from Detrital Zircons for the Existence of Continental Crust and Oceans on the Earth 4.4 Gyr Ago', Nature, CDIX (2001), pp. 175–8.

6 Kettanah et al., 'Provenance of the Ordovician–Lower Silurian Tumblagooda Sandstone, Western Australia'.

7 John W. Wells, 'Coral Growth and Geochronometry', Nature, CXCVI (1963), pp. 948–50.

8   Kenneth J. McNamara, 'Early Paleozoic Colonisation of the Land – Evidence from the Tumblagooda Sandstone, Southern Carnarvon Basin, Western Australia', *Journal of the Royal Society of Western Australia*, xcvii (2014), pp. 111–32.

9   Nigel H. Trewin and K. J. McNamara, 'Arthropods Invade the Land: Trace Fossils and Palaeoenvironments of the Tumblagooda Sandstone (?Late Silurian) of Kalbarri, Western Australia', *Transactions of the Royal Society of Edinburgh*, lxxxv (1995), pp. 177–210.

10 Margaret A. Bradshaw, 'Paleoenvironmental Interpretations and Systematics of Devonian Trace Fossils from the Taylor Group (Lower Beacon Supergroup), Antarctica', *New Zealand Journal of Geology and Geophysics*, xxiv/5–6 (1981), pp. 615–52.

11 Jim O. Buckman, '*Heimdallia* from the Lower Carboniferous of Ireland: *H. mullaghmori* a New Ichnospecies, and Re-Evaluation of the Three-Dimensional Format of the Ichnogenus', *Ichnos*, v/1 (1996), pp. 43–51.

12 Trewin and McNamara, 'Arthropods Invade the Land'.

13 Phillips et al., 'Further Information Relating to "A Geological Perspective on the Stone of Destiny".

7 Turn of the Seasons

1   Peter Worsley, 'Geology of the Clatford Bottom Catchment and Its Sarsen Stones on the Marlborough Downs', *Mercian Geologist*, xix/4 (2019), pp. 242–52.

2   David J. Nash et al., 'Origins of the Sarsen Megaliths at Stonehenge', *Science Advances*, vi/31 (2020), accessed at www.science.org.

3   Michael A. Summerfield and A. S. Goudie, 'The Sarsens of Southern England: Their Palaeoenvironmental Interpretation with Reference to Other Silcretes', in *The Shaping of Southern England*, ed. D.K.C. Jones (London, 1980), pp. 71–100.

4   Andy King and P. Collins, *A Building Stone Atlas of Berkshire* (Keyworth, 2017).

5   Charles E. Long, ed., *Diary of the Marches of the Royal Army During the Great Civil War; Kept by Richard Symonds, Now First Published from the Original ms. in the British Museum* (London 1859), p. 151.

6   Ibid.

7   Worthington G. Smith, *Man, the Primeval Savage* (London, 1894), p. 74.

8   Chris M. Green, 'Silcretes (Sarsen and Puddingstone) in England and Normandy since Stonehenge', *Proceedings of the Geologists' Association*, cxxvii/3 (2016), pp. 349–58.

9   John Hopkinson, 'Excursion to Radlett', *Proceedings of the Geologists' Association*, xi (1884), p. 453.

10 Green, 'Silcretes (Sarsen and Puddingstone) in England and Normandy Since Stonehenge'.

11 Ibid., pp. 355–6.

12 Jane Tubb, 'Palaeogene Conglomerates (Puddingstones) in the Colliers End Outlier, East Hertfordshire, UK – Evidence for Silicification in the Basal Reading Formation', *Proceedings of the Geologists' Association*, CXXVII (2016), pp. 320–26.

13 W. J. Eric van de Graaff, 'Silcrete in Western Australia: Geomorphological Settings, Textures, Structures, and Their Genetic Implications', in *Residual Deposits: Surface Related Weathering Processes and Materials*, ed. R.C.L. Wilson (London, 1983), pp. 159–67; J. S. Ullyott et al., 'Distribution, Petrology and Mode of Development of Silcretes (Sarsens and Puddingstones) on the Eastern South Downs, UK', *Earth Surface Processes and Landforms*, XXIX (2004), pp. 1509–39; M. Thiry and A. R. Milnes, 'Silcretes: Insights into Occurrences and Formation of Materials Sourced for Stone Tool Making', *Journal of Archaeological Science: Reports*, XV (2017), pp. 500–513.

14 Thomas Westerhold et al., 'An Astronomically Dated Record of Earth's Climate and Its Predictability over the Last 66 Million Years', *Nature*, CCCLXIX/6509 (2020), pp. 1383–7.

15 Jussi Hovikoski et al., 'Paleocene–Eocene Volcanic Segmentation of the Norwegian–Greenland Seaway Reorganized High-Latitude Ocean Circulation', *Communications Earth and Environment*, II/172 (2021), p. 172, accessed at www.nature.com.

16 Wie Wang et al., 'Tectonic and Climatic Controls on Sediment Transport to the Southeast Indian Ocean during the Eocene: New Insights from IODP Site U1514', *Global and Planetary Change*, CCXII (2022).

17 Paul R. Gammon et al., 'Sedimentology and Lithostratigraphy of Upper Eocene Sponge-Rich Sediments, Southern Western Australia', *Australian Journal of Earth Sciences*, XLVII/6 (2000), pp. 1087–103.

18 Charles R. M. Butt, 'Granite Weathering and Silcrete Formation on the Yilgarn Block, Western Australia', *Australian Journal of Earth Sciences*, XXXII/4 (1985), pp. 415–32.

19 Van der Graaff, 'Silcrete in Western Australia: Geomorphological Settings, Textures, Structures, and Their Genetic Implications'.

20 Robert S. Hill, 'Fossil Evidence for the Onset of Xeromorphy and Scleromorphy in Australian Proteaceae', *Australian Systematic Botany*, XI/3–4 (1998), pp. 391–400.

21 Timothy Bata, 'Widespread Development of Silcrete in the Cretaceous and Evolution of the Poaceae Family of Grass Plants', *Earth Science Research*, V (2016), pp. 1–19.

22 Douglas W. Kirkland, 'An Explanation for the Varves of the Castile Evaporites (Upper Permian), Texas and New Mexico, USA', *Sedimentology*, L/5 (2003), pp. 899–920.

23 Ibid.

24 Morgane Ledevin et al., 'The Sedimentary Origin of Black and White Banded Cherts of the Buck Reef, Barberton, South Africa', *Geosciences*, ix/10 (2019), p. 424, accessed at www.mdpi.com.

25 Melvyn J. Lintern, 'The Association of Gold with Calcrete', *Ore Geology Reviews*, lxvi (2015), p. 136.

26 Melvyn J. Lintern et al., 'Ionic Gold in Calcrete Revealed by la-icp-ms, sxrf and xanes', *Geochemica et Cosmochinica Acta*, lxxiii/6 (2009), p. 5.

27 Ken McNamara, *Pinnacles* (Perth, 2009).

28 Matej Lipar and J. A. Webb, 'The Formation of the Pinnacle Karsts in Pleistocene Aeolian Calcarenites (Tamala Limestone) in Southwestern Australia', *Earth-Science Reviews*, cxl (2015), pp. 182–202.

29 Matej Lipar et al., 'Aeolianite, Calcrete/Microbialite and Karst in South-Western Australia as Indicators of Middle to Late Quaternary Palaeoclimates', *Palaeogeography, Palaeoclimatology, Palaeoecology*, cccclxx (2017), pp. 11–29.

30 Eric Verrecchia, 'Incidence de l'activité fungique sur l'induration des profils carbonatés de type calcrete pédologique. L'exemple du cycle oxalate–carbonate de calcium dans les encroûtements calcaires de Galilée (Israël)', *Compte rendus de l'Académie des sciences*, cccxi (1990), pp. 1367–74.

31 Lintern et al., 'Ionic Gold in Calcrete', pp. 1666–83.

32 Ibid., p. 1667.

33 Melvyn J. Lintern et al., 'Natural Gold Particles in *Eucalyptus* Leaves and Their Relevance to Exploration for Buried Gold Deposits', *Nature Communications*, iv/1 (2015), accessed at www.nature.com.

34 Matej Lipar et al., 'The Genesis of Solution Pipes: Evidence From the Middle–Late Pleistocene Bridgewater Formation Calcarenite, Southeastern Australia', *Geomorphology*, ccxlvi (2015), pp. 90–103.

8 Centres of Attention

1 Ravi R. Anand and M. Verrall, 'Biological Origin of Minerals in Pisoliths in the Darling Range of Western Australia', *Australian Journal of Earth Sciences*, lviii/7 (2011), pp. 823–33.

2 Ibid., p. 823.

3 Ibid.

4 Martin A. Wells et al., '(u-th)/he-Dating of Ferruginous Duricrust: Insight into Laterite Formation at Boddington, wa', *Chemical Geology*, dxxii (2019), pp. 148–61.

5 Anand and Verrall, 'Biological Origin of Minerals in Pisoliths'.

6 Wells et al., '(u-th)/he-Dating of Ferruginous Duricrust: Insight into Laterite Formation at Boddington, wa'.

7   John M. Keynes, 'Newton, the Man', in Royal Society, *Newton Tercentenary Celebrations* (Cambridge, 1947), pp. 27–34.

8   John Woodward, *An Addition to the Catalogue of the Foreign Native Fossils, in the Collection of J. Woodward* MD (London, 1728), specimen 0.1; see S. Buckelow, *Red: The Art and Science of a Colour* (London, 2016), p. 71.

9   Leonard W. Johnson and M. L. Wolbarsht, 'Mercury Poisoning: A Probable Cause of Isaac Newton's Physical and Mental Ills', *Notes and Records of the Royal Society of London*, XXXIV/1 (1979), pp. 1–9.

10 John M. Stillman, 'Paracelsus as a Chemist and Reformer of Chemistry', *The Monist*, XXIX (1919), p. 115.

11 John Woodward, *An Attempt Towards a Natural History of the Fossils of England; in a Catalogue of the English Fossils in the Collection of J. Woodward,* MD *Containing a Description and Historical Account of Each; with Observations and Experiments, Made in Order to Discover, as Well the Origin and Nature of Them, as Their Medicinal, Mechanical, and Other Uses, Tome 1* (London, 1729), p. 84.

12 Ana M. Alfonso-Goldfarb et al., 'Seventeenth-Century "Treasure" Found in Royal Society Archives: The *Ludus Helmontii* and the Stone Disease', *Notes and Records of the Royal Society*, LXVIII (2014), pp. 227–43.

13 Ibid., p. 235.

14 Woodward, *An Attempt Towards a Natural History of the Fossils of England*, pp. 81–2.

15 Ibid., p. 82.

16 Gideon A. Mantell, 'Notice of the Remains of *Dinornis* and Other Birds, and of Fossils and Rock-Specimens, Recently Collected by Mr. Walter Mantell in the Middle Island of New Zealand', *Proceedings of the Geological Society*, VI (1850), pp. 319–42.

17 James R. Boles et al., 'The Moeraki Boulders: Anatomy of Some Septarian Concretions', *Journal of Sedimentary Petrology*, LV/3 (1985), pp. 398–406.

18 Brian R. Pratt, 'Septarian Concretions: Internal Cracking Caused by Synsedimentary Earthquakes', *Sedimentology*, XLVIII/1 (2001) pp. 189–213.

19 James P. Hendry et al., 'Jurassic Septarian Concretions from NW Scotland Record Interdependent Bacterial, Physical and Chemical Processes of Marine Mudrock Diagenesis', *Sedimentology*, LIII/3 (2006), pp. 537–65.

20 John K. Wright, 'The Early Kimmeridgian Ammonite Succession at Staffin, Isle of Skye', *Scottish Journal of Geology*, XXV/3 (1989), pp. 263–72.

21 Hendry et al., 'Jurassic Septarian Concretions', p. 540.

22 Ibid., p. 561.

23 Ibid.

24 John A. Long and K. Trinajstic, 'The Late Devonian Gogo Formation Lägerstatte of Western Australia: Exceptional Early Vertebrate Preservation and Diversity', *Annual Review of Earth and Planetary Sciences*, XXXVIII (2010), pp. 255–79.

25 Kate Trinajstic et al., 'Exceptional Preservation of Nerve and Muscle Tissue in Late Devonian Placoderm Fish and Their Evolutionary Implications', *Biology Letters*, III/2 (2007), pp. 197–200.

26 John A. Long et al., 'Live Birth in the Devonian Period', *Nature*, CDLIII/7195 (2008), pp. 650–52.

27 Hidekazu Yoshida et al., 'Generalized Conditions of Spherical Carbonate Concretion Formation around Decaying Organic Matter in Early Diagenesis', *Scientific Reports*, VIII/6308 (2018), pp. 1–10.

28 A Lagerstätte is an exceptionally preserved fossil deposit, frequently, though not always, of both hard parts and soft tissue.

29 Thomas Clements et al., 'The Mazon Creek Lagerstätte: A Diverse Late Paleozoic Ecosystem Entombed within Siderite Concretions', *Journal of the Geological Society*, CLXXVI (2019), pp. 1–11.

30 Ibid.

31 Ibid.

32 B. Woodland and R. Stenstrom, 'The Occurrence and Origin of Siderite Con-cretions in the Francis Creek Shale (Pennsylvanian) of Northeastern Illinois', in *Mazon Creek Fossils*, ed. M. H. Nitecki (New York, 1979), pp. 69–103.

9 FOSSILS, FORESTS AND FIRE

1 Kenneth J. McNamara and R. Feist, 'The Effect of Environmental Changes on the Evolution and Extinction of Late Devonian Trilobites from the Northern Canning Basin, Western Australia', in *Devonian Climate, Sea-Level and Evolutionary Events*, ed. R. T. Becker et al. (London, 2016), pp. 251–71.

2 Ibid., p. 263.

3 Ibid., p. 265.

4 Simon B. Laughlin et al., 'The Metabolic Cost of Neural Information', *Nature Neuroscience*, I (1998), pp. 36–41.

5 Justin Merry et al., 'Variation in Compound Eye Structure: Effects of Diet and Family', *Evolution*, LXV/7 (2011), pp. 2098–110.

6 Thomas J. Algeo et al., 'Late Devonian Oceanic Anoxic Events and Biotic Crises: "Rooted" in the Evolution of Vascular Land Plants?', *GSA Today*, V/3 (1995), pp. 64–6.

7 William E. Stein et al., 'Mid-Devonian *Archaeopteris* Roots Signal Revolutionary Change in Earliest Fossil Forests', *Current Biology*, XXX (2020), pp. 421–31.

8 Dianne Edwards and L. Axe, 'Anatomical Evidence in the Detection of the Earliest Wildfire', *Palaios*, XIX (2004), pp. 113–28.

9 Kunio Kaiho et al., 'A Forest Fire and Soil Erosion Event during the Late Devonian Mass Extinction', *Palaeogeography, Palaeoclimatology, Palaeoecology*, CCCXCII (2013), pp. 272–80.

10   Matthew S. Smart et al., 'Enhanced Terrestrial Nutrient Release During the Devonian Emergence and Expansion of Forests: Evidence from Lacustrine Phosphorus and Geochemical Records', *Bulletin of the Geological Society of America* (2022), accessed at http://pubs.geoscienceworld.org.
11   McNamara and Feist, 'The Effect of Environmental Changes'.
12   The Reef 2050 Long-Term Sustainability Plan is the Australian and Queensland government's overarching framework for protecting and managing the Great Barrier Reef to 2050, available at www.environment.gov.au.
13   John Woodward, 'Some Thoughts and Experiments Concerning Vegetation', *Philosophical Transactions of the Royal Society*, XXI (1699) pp. 193–227.
14   Ibid., p. 208.
15   Richard H. Grove, 'A Historical Review of Early Institutional and Conservationist Responses to Fears of Artificially Induced Global Climate Change: The Deforestation–Desiccation Discourse 1500–1860', *Chemosphere*, XXIX/5 (1994), pp. 1001–13.
16   Woodward, 'Some Thoughts and Experiments Concerning Vegetation', p. 209.
17   Vivi Vajda et al., 'End-Permian (252 Mya) Deforestation, Wildfires and Flooding – An Ancient Biotic Crisis with Lessons for the Present', *Earth and Planetary Science Letters*, DXXIX (2020), accessed at www.sciencedirect.com.
18   Robert Scott's diary entry, 8 February 1912, accessed at spri.cam.ac.uk.
19   Stephen McLoughlin et al., 'Age and Paleoenvironmental Significance of the Frazer Beach Member – A New Lithostratigraphic Unit Overlying the End-Permian Extinction Horizon in the Sydney Basin, Australia', *Frontiers in Earth Science*, VIII (2021), pp. 1–31, accessed at www.frontiersin.org.
20   Daoliang Chu et al., 'Ecological Disturbance in Tropical Peatlands Prior to Marine Permian–Triassic Mass Extinction', *Geology*, XLVIII/3 (2020), accessed at https://pubs.geoscienceworld.org.
21   Chris Mays and S. McLaughlin, 'End-Permian Burnout: The Role of Permian–Triassic Wildfires in Extinction, Carbon Cycling and Environmental Change in Eastern Gondwana', *Palaios*, XXXVII (2022), pp. 292–317.
22   Ibid.
23   McLoughlin et al., 'Age and Paleoenvironmental Significance of the Frazer Beach Member', p. 25.
24   Craig Woodward et al., 'The Hydrological Legacy of Deforestation on Global Wetlands', *Science*, CCCXLVI/6211 (2014), pp. 844–7.
25   McLoughlin et al., 'Age and Paleoenvironmental Significance of the Frazer Beach Member', p. 25.
26   Ibid., p. 26.
27   Grove, 'A Historical Review of Early Institutional and Conservationist Responses to Fears of Artificially Induced Global Climate Change', p. 1002.
28   Ibid.

29 Mark Andrich and J. Imberger, 'The Effect of Land Clearing on Rainfall and Fresh Water Resources in Western Australia: A Multi-Functional Sustainability Analysis', *International Journal of Sustainable Development and World Ecology*, xx/6 (2013), pp. 549–63.

30 Ibid.

BIBLIOGRAPHY

Beerling, David, *The Emerald Planet: How Plants Changed Earth's History* (Oxford, 2007)

Fortey, Richard, *Earth: An Intimate History* (New York, 2004)

——, *Life: A Natural History of the First Four Billion Years of Life on Earth* (New York, 1997)

Friend, Peter, *Southern England: Looking at the Natural Landscapes* (London, 2008)

James, Noel P., and Brian Jones, *Origin of Carbonate Sedimentary Rocks* (Chichester, 2018)

Knoll, Andrew H., *Life on a Young Planet: The First Three Billion Years of Evolution on Earth* (Princeton, NJ, 2015)

Konhauser, Kurt, *Introduction to Geomicrobiology* (Oxford, 2007)

MacDougall, Doug, *Why Geology Matters* (Oakland, CA, 2011)

McNamara, Ken, 'Early Paleozoic Colonisation of the Land: Evidence from the Tumblagooda Sandstone, Southern Carnarvon Basin, Western Australia', *Journal of the Royal Society of Western Australia*, XCVII (2014), pp. 111–32

——, 'Earth and Life: Origins of Phanerozoic Diversity', *Australian Journal of Earth Sciences*, LV (2008), pp. 1023–36

——, *Pinnacles* (Perth, 2009)

——, *Stromatolites* (Perth, 2009)

——, and John Long, *The Evolution Revolution: Design without Intelligence* (Melbourne, 2007)

Oxford University Museum of Natural History, ed., *Strata: William Smith's Geological Maps* (London, 2020)

Porter, Roy, *The Making of Geology: Earth Science in Britain, 1660–1815* (Cambridge, 1977)

Stow, Dorrik A.V., *Sedimentary Rocks in the Field: A Colour Guide* (Boca Raton, FL, 2005)

Walkden, Gordon, *Devonshire Marbles: Their Geology, History and Uses*, 2 vols (London, 2015)

Woodcock, Nigel, and Rob Strahan, eds, *Geological History of Britain and Ireland* (Chichester, 2012)

ACKNOWLEDGEMENTS

Quarries offer not only vivid insights into the underworld but, on occasion, chance meetings that can change the course of our lives. The greater part of my journey investigating rocks, and particularly their organic remains, I have shared with my wife, Sue Radford. Appropriately, we met in a Carboniferous limestone quarry in Dunbar, Scotland. Without her support and encouragement over many years (both inside and outside of quarries) it is unlikely that I would ever have been able to write this book. Her perceptive advice on many aspects of the manuscript and her wealth of lithological insights have been invaluable. I offer her my sincere thanks.

I also wish to thank Jamie McNamara, Mike Green and Tony Marrion for reading parts or all of the manuscript and providing welcome feedback. One of the delights in writing a book such as this is the unwavering support of colleagues. This is especially so in the provision of their photographs or images in their care, as well as pointing me in the direction of new discoveries. Invariably the responses to my requests have been rapid and positive, and their kindness has done much to improve this book. In particular, I am most grateful to Wladyslaw Altermann, Stephen Anstey, Christopher Berry, Di Brooks, Jeff Brooks, Stan Celestian, Michael Chazan, Renee Doropoulos, Peter Downes, Chris Fielding, Tracy Frank, Christopher Green, Sarah Hammond, Steve McLoughlin, Peter Pope, William Stein, Donna Wajon, Eddy Wajon and Gordon Walkden.

I am most grateful for the support of Michael Leaman at Reaktion Books, particularly for encouraging me to write this book. I also wish to thank Amy Salter for her expert editorial endeavours and Susannah Jayes for picture editing.

PHOTO ACKNOWLEDGEMENTS

The author and publishers wish to express their thanks to the below sources of illustrative material and/or permission to reproduce it.

Wladyslaw Altermann: p. 139; Jeff and Di Brooks: p. 251; Stan Celestian: pp. 85, 197; Michael Chazan: p. 150; image courtesy of the Geological Survey and Resource Strategy, Department of Mines, Industry Regulation and Safety © State of Western Australia 2021: p. 46; Christopher Green: p. 185; Historic Environment Scotland: p. 156; Steve McLoughlin, Chris Fielding and Tracy Frank: p. 262; Ken McNamara: pp. 22, 27, 33, 38, 52, 54, 57, 61, 78, 81, 88, 96, 102, 113, 120, 123, 125, 133, 141, 159, 167 top and bottom, 171, 174, 178, 183, 189, 193, 201, 204, 207, 209, 213, 217, 220, 231, 240, 242 top and bottom, 244, 259; Linda Moore: p. 41 top and bottom; Public Domain: pp. 10, 11, 69, 94, 107, 117, 153; Sedgwick Museum of Earth Sciences, Cambridge: pp. 15, 16, 19, 234; William Stein and Christopher Berry: p. 247, Gordon Walkden: p. 92; Western Australian Museum, Perth: pp. 87, 136.

Wilson44691 (Mark A. Wilson, Department of Geology, The College of Wooster), the copyright holder of the image on p. 71, has released this work into the public domain. Nachoman-au, the copyright holder of the image on p. 99, has published it online under conditions imposed by a Creative Commons Attribution-Share Alike 3.0 Unported License. ja:User:NEON / commons:User:NEON_ja, the copyright holder of the image on p. 105, Keith Pomakis, the copyright holder of the image on p. 224, and Kevin Prince, the copyright holder of the image on p. 226, have published the images online under conditions imposed by a Creative Commons Attribution-Share Alike 2.5 Generic License. James St John, the copyright holder of the image on p. 163, has published it online under conditions imposed by a Creative Commons Attribution 2.0 Generic License.

INDEX

Page numbers in *italics* refer to illustrations